THIRD WA PROJECT MANAGEMENT

A Handbook for Managing the Complex Information Systems for the 1990s

Rob Thomsett

Yourdon Press Computing Series
Prentice Hall
Englewood Cliffs, New Jersey 07632

Library of Congress Cataloging-in-Publication Data

Thomsett, Rob.
Third wave project management : a handbook for managing the complex information systems of the 1990s / Rob Thomsett.
p. cm.
Includes bibliographical references and index.
ISBN 0–13–915299–7
1. Information resources management. I. Title.
T58.64.T46 1993
004.2'1'0684–dc20 92–11023
CIP

Line art of workers as shown in R.S.V.P. #17, *The Directory of Illustration and Design,* by Jud Guitteau. Courtesy of Renard Represents, Inc.

Acquisitions editor: Paul Becker
Production and interior design: Fred Dahl
Cover design: Rick Dombrowski
Prepress buyer: Mary McCartney
Manufacturing buyer: Susan Brunke / Mary McCartney

Printed in the United States of America

10 9 8 7 6 5 4 3 2 1

ISBN 0-13-915299-7

Prentice-Hall International (UK) Limited, *London*
Prentice-Hall of Australia Pty. Limited, *Sydney*
Prentice-Hall Canada Inc., *Toronto*
Prentice-Hall Hispanoamericana, S.A., *Mexico*
Prentice-Hall of India Private Limited, *New Delhi*
Prentice-Hall of Japan, Inc., *Tokyo*
Prentice-Hall of Southeast Asia Pte. Ltd., *Singapore*
Editora Prentice-Hall do Brasil, Ltda., *Rio de Janeiro*

A special thanks to the following battle veterans who reviewed and improved earlier drafts of this book:

John Wright, Computer Sciences of Australia

John Van Waterschoot, Department of Defence

Keith Parrott, Australian Archives

Contents

Introduction: Back to the Future

This book is about a new approach to project management. This approach reflects the massive changes in systems, technology, and development techniques that have arrived during the 1980s with the same unpredictability and urgency as the financial and social changes that have altered the business environment forever.

In 1980, I wrote *People and Project Management,* which summarized the learnings I had gained from working with over 1,000 experienced project managers in the 1970s. In that book, reiterated Joel Aron's observation, made in 1969, that "we ran into trouble because we didn't know how to manage what we had, not because we lacked the techniques." In other words, what we had learned was that the key to success in information systems development was in managing people, not technology.

I also argued that computer people deserved the right to be managed in a way that recognized their need to develop quality systems in a professional manner. I also suggested that the simple application of well-established, people-oriented project management and team techniques would unlock a tremendous potential for productivity and quality, as well as creating a "fun" environment for people to work in.

Clearly, as I write this handbook in the 1990s, to paraphrase Warren Bennis, "Something funny happened on the way to the future." The "excesses" of the 1980s (a term widely used in the business press when discussing the past decade) have created another generation of computer people who have survived a possibly even more oppressive environment than that of their peers in the 1970s.

Toward the end of the 1970s, we believed that the social and cultural revolution begun in the 1960s had led to organizations' questioning their management practices and, in particular, the long-established hierarchical and bureaucratic principles prevalent in all major organizations with their associated dysfunctional human effects of alienation, low morale, and productivity.

We believed that the 1980s would see a fundamental shift toward organizations that did not see any inconsistency between the productivity and the empowerment of their people.

We believed that people and project management would be vital elements of such a shift.

We Interrupt This Program for . . .

Instead, the 1980s faced unprecedented and unforeseen challenges in globalization, competition, technological change, deregulation, and growth, which change the rules under which all organizations operated. In particular, computing was confronted by two clearly identified pressures: product and system proliferation, and the emergence of a third wave of computing technology and technique.

The increasing pressure from competition, global markets, and government regulation and deregulation required organizations to respond through new products, new alliances, and new markets. In particular, this turbulence in the business environment required computing groups to respond to demands for new products and enhancements to existing products in an era of fixed deadlines and market-driven constraints. However, as discussed later, most computing groups were ill equipped for this, having become complacent in their existing roles and technology.

Rather than examining these pressures from a long-term perspective, most organizations met these challenges by simply throwing more people and more technology at the problem. Companies with which my group worked increased the size of their computing groups by an order of magnitude! Further, the quality of software and, in some cases, the quality of computing skills were compromised in the rush to respond to the demands being placed on computing groups for new systems and technology. By the mid-1980s, shortages of skilled computer people emerged as a national concern, and highly skilled computer people became the focus of spiralling wage increases and high turnover. Throughout the late 1980s, turnover rates in computing averaged 35 percent per year.

As shown in Figure I.1, the conjunction of these two pressures simply placed computer groups into a poorly planned, reactive mode. In the figure, the *y*-axis is the number of retail banking products (check, savings, plastic, loan, etc.) offered to the Australian public by a leading bank, and the *x*-axis models the evolution of computing technology and techniques over the past 30 years.

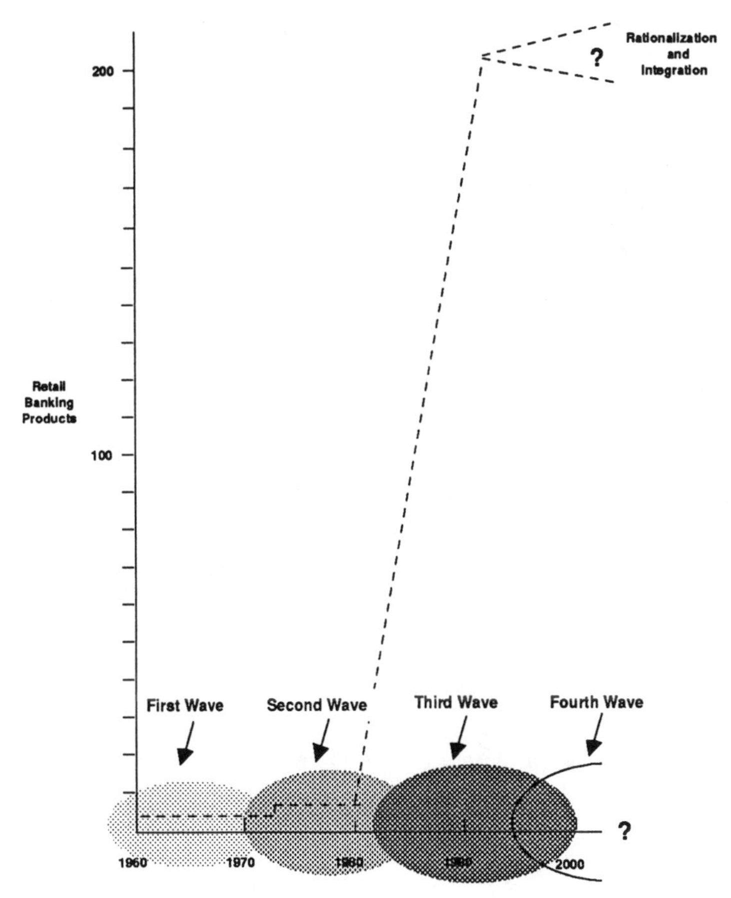

Figure I.1 Product Innovation and New Technology

The Third Wave of Computing

The evolution of commercial computing technology and technique can be traced from the early 1960s. The initial development of computing technology from the 1950s really was focused in the defense, aerospace, and scientific research laboratories and, as such, is ignored in this model.

The first wave lasted through the 1960s and was typified by batch mainframes, COBOL and FORTRAN, unstructured and "entrepreneurial" system development techniques, automation of core systems, and nonexistent project management and client/user involvement. In effect, the newly emerged computing experts dominated the process and were left alone to practice their "dark art."

The second wave was, in part, a reaction to the first wave, which left a legacy of low quality systems, but was also driven by new technology such as hierarchical data bases and rudimentary networks. The second wave was marked by the structured revolution where developments in process modeling (data flow diagrams), structured design, and data analysis from people such as Larry Constantine & Ed Yourdon [1979], Clive Finklestein [1989], and Chris Gane and Trish Sarson [1979] provided a new and more engineered approach to system development. Developments in data base technology provided an opportunity for the rewriting of many of the first wave systems and the easier interfacing of previously unintegrated systems. Highly rigid project management techniques, borrowed for the construction and manufacturing professions, were implemented and there was the "token" involvement of clients/users in the systems specification and testing activities.

The third wave began to emerge in the late 1970s and confronted many other existing practices now firmly entrenched in computing groups. The rapid growth of personal computers was a key element of the third wave. The dominance of relational data bases such as IBM's DB2 and the integration of the previously "warring" factions of the data and functional modellers into Information Engineering led to the development of integrated system architectures and the commencement of megaprojects to rewrite (again!) the semi-integrated systems from the second wave into completely integrated systems. (These were typically based around a small set of key data supertypes, such as client and product.)

However, probably the most significant developments in the third wave were not directly associated with technology or technique. First, many clients and users—frustrated with a lack of perceived service from computing people and reacting to unprecedented demands for new products, services, and competitive response—began to gain control of computing priorities, budgets, and direction. In addition, traditional project management techniques were completely inadequate when faced with the new regime of fixed deadlines, dynamic user requirements, and high people turnover.

The Need for a New Project Management Model

As discussed throughout this handbook, the trend to place computing groups on a commercialized cost-recovery or profit-center basis often found traditional computing project management wanting. Further, the emergence of new development techniques such as Joint Requirements Planning, Joint Application Design, Rapid Application Development, time-boxing [Martin, 1990] and Fast-Tracking required a more dynamic and real-time project management approach than the one suited to the more "opulent" development approaches of the second wave.

As the third wave entered the 1990s, there emerged some developments that may be considered as the harbingers of a fourth wave. Downsizing development to workstations [Yourdon and Nash, 1991], object-oriented development [Yourdon and Nash, 1989], and IBM's AD/Cycle [Mercurio et al., 1990] are considered by some observers as fourth wave components.

Rather than be caught up in the evolutionary versus revolutionary debate currently centered around the object-oriented paradigm, I will simply "sit on the fence" and, as shown in Figure I.2, suggest that the issues of project management and team management are similar for these new technologies as those associated with the third wave. In other words, the third and fourth wave technologies and techniques demand a new project management paradigm.

In addition, as discussed by Thomsett [1992] and Constantine [1989], they require a new focus on team formation, structure, and management. The pressures for increased productivity, fewer people and more client-oriented service must change the typically ad hoc approach to selecting project team members and the "team for life" concept prevailing in most computing organizations.

Now Back to the Regular Programs . . .

There are signs that many organizations have finally realized that the people-oriented values developed by many forces in the 1960s and 1970s are fundamentally sound and that outmoded organization structures sustained through the 1980s by the "throwing-money-at-it" syndrome will be meaningfully restructured to reflect these values.

The one difference now is that, whereas in the 1970s the forces for change were founded in a belief of human values and dignity, the forces for change in the 1990s are based on perhaps a much purer principle—survival.

I have taught project management to over 5,000 computer and business professionals in the United States, Australia, the United Kingdom, and Hong Kong over the past decade. This handbook is dedicated to those people and is written in the hope that they will finally experience the productive, quality, human, and fun work environment that I wished for the people who are now their bosses.

A fundamental hope that we all have for the 1990s is professional third wave project management that:

- Recognizes the imperatives of the current business environment.
- Yet also recognizes that every computer and business professional has the right to work on a well managed project.
- Treats them as thinking human beings who choose freely to work on projects that improve both their existence and their organizations'.

This book provides a detailed guide to this new model of project management.

Rob Thomsett
Canberra

	First Wave	Second Wave	Third Wave (Third Wave)	Fourth Wave? (Third Wave)
Methodologies	Unstructured	Structured (Process), Data Analysis	Information Engineering (integration of data & function), RAD	Object-oriented, integration with Expert CASE
Information Systems	Stand-alone core systems	Core systems rewritten & support systems	Core systems, support systems, entrepreneurial systems	Entrepreneurial systems
Information System Architectures	Non-integrated	Interfaced	Semi-integrated	Fully integrated
Languages/ Development Technology	Assembler 3 GLs (COBOL, PL/1)	New 3 GLS (eg C, Ada, etc), 4GLs (Natural, Mapper, etc)	Application Generators, ICASE, PC-languages	Object-oriented Languages (C++, Eiffel), Expert CASE
Databases	Hierarchical, flat file processing	Network databases	Relational databases	Object-oriented, Distributed databases
Networks	Point-to-point, simple protocols	Sophisticated proprietary protocols eg SNA		Integrated voice and data networks, cordless connection
Workstations	Teletype	Block-mode 3270 style	PCs as 3270 emulation and intelligent terminals	PCs fully integrated as development technology
Processors (O.S.)	Main-frames (Batch, time-sharing)	Mini's (Multi-tasking eg MVS and VMS)	PCs, specialized processors (Unix, Windows, OS/2)	Highly scalable and parallel architectures (new mainframe OS?)

Fig. I.2 The Three Waves of Computing

References

J. D. Aron, ''The Super-Programmer Project,'' Software Engineering Techniques: Report of the 1969 Rome Conference, J.N. Buxton and B. Randell, eds. Brussels, Belgium: NATO Science Committee, 1970.

L. L. Constantine, ''Team Paradigms and the Structured Open Team,'' Proceedings of the Embedded Systems Conference, San Francisco, September 1989.

C. Finklestein, *An Introduction to Information Engineering*. Reading, Mass.: Addison-Wesley, 1989.

C. Gane & T. Sarson, *Structured Systems Analysis: Tools and Techniques*. Englewood Cliffs, N.J.: Prentice-Hall, 1979.

V. F. Mercurio, B. F. Meyers, A. M. Nisbet & G. Radin, ''AD/Cycle strategy and architecture,'' *IBM Systems Journal,* Vol. 29, No. 2, 1990, pp. 170–187.

R. Thomsett, ''The X-team: A new project team structure,'' *American Programmer,* Vol. 5, No. 1, January 1992.

E. N. Yourdon & L. L. Constantine, *Structured Design*. Englewood Cliffs, N.J.: Prentice-Hall, 1979.

E. N. Yourdon & T. Nash (eds.), ''Downsizing,'' *American Programmer,* Vol. 4, No. 8, August 1991.

E. N. Yourdon & T. Nash (eds.), ''Object-oriented Observations,'' *American Programmer,* Vol. 2, Nos. 7–8, Summer 1989.

1 New Project Management Concepts

A funny thing happened on the way to the future.
Warren Bennis [1970]

Project management is a creative problem-solving process.

It is an essential management process that can make the difference between a project's failure or success.

The approach to project management outlined in this handbook incorporates a number of philosophies that are missing or incomplete in other project management approaches. The conceptual framework of the project management approach has been developed, implemented, and revised over the past 12 years in a number of U.S., Asian, and Australian organizations by the Thomsett Associates consulting team. While the approach is simple, it integrates a number of powerful concepts.

Reflecting the turbulent business and social environment of the 1980s and 1990s, it recognizes the need for projects to be developed in a time-critical and cost-sensitive manner and incorporates the prevailing management principles of participative team work and client control of information systems.

Project Management Concepts

There are a number of critical concepts in the frame of mind underlying this successful project management approach:

- Managerial versus technical control.
- Project management as a problem-solving process.
- Team-driven project management.
- Senior management involvement.
- Rapid Application Planning sessions.
- The critical project management information set.
- Real-time project management.

Managerial Versus Technical Control

Project management is about the managerial and business aspects of a project. Such things as costs, benefits, risks, people, deadlines, deliverables, contingencies, productivity, and priorities are the stuff of project management.

A software development and enhancement project can be defined using two related sets of information: the management or business set and the technical set. As shown in Figure 1.1 these two sets are joined by the objectives, scope, quality, and strategy of the project. The technical issues related to a project must be resolved by the technical people and the relevant technical control system. For example, problems—such as the software has defects, or the network is working, or the policy for staffing has legal issues involved—are technical concerns and should be solved by the most competent technical people.

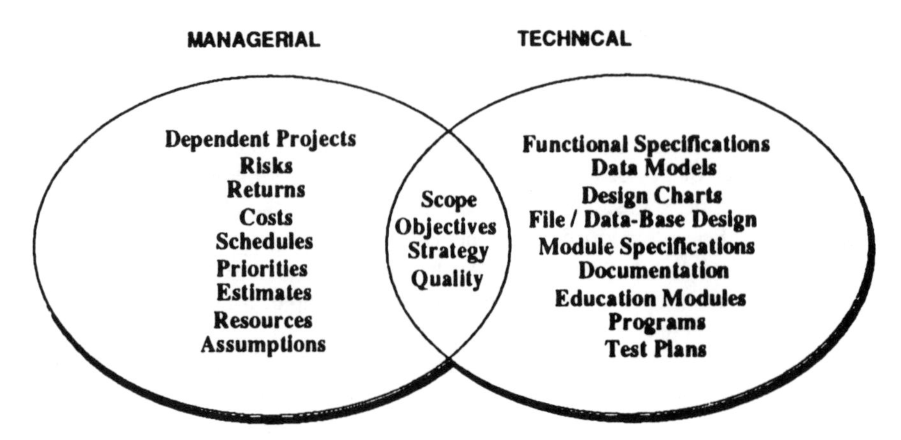

Fig. 1.1
Project Definition—Management and Technical

The problem the project management process must solve is the *impact* of the technical issues on the project management concerns such as deadlines, deliverables, people, and so on. For example, if the software has defects, then the project manager must determine what he or she can do in terms of resources, money, alteration of effort, or priority to enable the technical solution to be implemented. The project manager *facilitates* the process of technical solution but doesn't get involved in the problem solution.

The common information of scope, objectives, strategy, and quality (requirements) are of concern both to project managers and to technical managers and experts. This information provides the linking pin between the managerial and technical issues of a project.

Simply, the project management process is interrelated with the technical nature of the project, but it should be as independent as possible from the detailed technical issues.

Project Management as a Problem-Solving Process

It is a reasonable assumption that all projects will be subject to change, disruptions, and stress.

As shown in Figure 1.2, detailed studies have revealed that the typical project will involve alternation of objectives and expansion of scope, with a resulting poor estimation at the initial states of the project. In addition, many projects will be subject to staff turnover, quality problems, changing of priorities, and other issues related to the long duration of some projects. As a result, project management involves the monitoring, evaluation, and control of the variances to the initial project concepts.

The project management process must focus during the *planning activity*

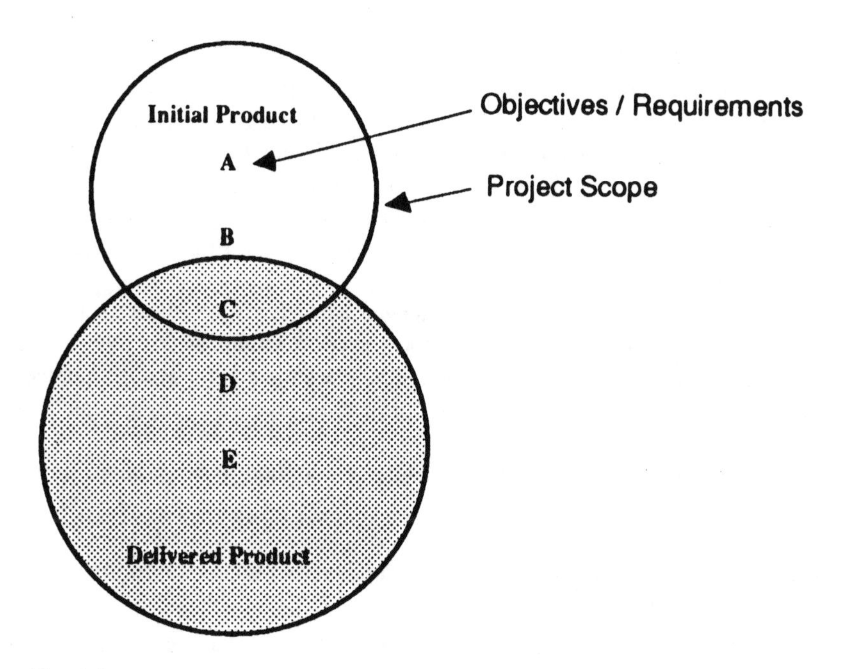

Fig. 1.2
Project Variance

on the factors that are likely to cause project variance. An effective project management process seeks to predict and reduce the probability of project variation *before* it occurs and to have prepared contingency plans for the inevitable project disruptions.

In many ways, the concept of project management as a problem-solving process underlies the emergence of contemporary management practice, which has been driven over the past decade by people such as Peters [1988], Drucker [1985], Handy [1989], and many others.

Many projects spend more time and effort in a reactive process of getting out of project disasters than they would be prepared to spend in proactive planning to avoid those disasters. There is a mind set in many technical people and technical managers that planning is an overhead and that it diverts them from their *real* work which is design, programming, and so on.

It should be emphasized that project planning and project management are not overheads but rather normal work activities carried out as part of the project team's daily activities.

Identification of potential problems in a project, along with proactive resolution or minimization of these problems, is a key element of the project management approach developed by Thomsett Associates.

Team-Driven Project Management

As for all problem-solving approaches, the use of team-based techniques and formal problem-oriented methods are integral to effective project management. Again, the use of team-based processes as distinct from the classic ''lone'' project manager process of traditional project management reflects the implementation of these techniques in other industries pioneered by Taiichi Ohno of Toyota [Womack, Jones & Roos, 1990], Fred Emery and Eric Trist [1965], and other organization reformers.

It is essential that the project management activity be undertaken by as many team members, project users, and key support groups as possible. Should the project manager be in a situation of planning a project without team members available, he or she should involve other project managers and key groups in the process.

A number of advantages result from the adoption of team-based project management:

- The project plans, supporting information and project tracking information will be more accurate.
- The team members will have a commitment to the plans as the team was integral to the development of the plans.
- The project information will be more politically acceptable because of the broad base of input.
- It is more fun!

The team-based project planning and management approach is a highly disciplined one. Throughout the handbook, various hints and guidelines are given to assist in structuring and controlling the team planning sessions.

Senior Management Involvement

The role of senior management is critical in Thomsett Associates' approach to project management.

In too many organizations, senior managers are placed on Steering Committees with little education and support to enable them to fully participate in managing the project. They are often sidetracked into detailed technical issues or presented with problems *after* they have occurred which place them in a reactive mode.

Senior management must be treated as the high-level project management and problem-solving group in a project. Whether they are called Steering Committees, Project Review Committees, or simply ''them,'' senior managers are a vital *support* network for the project manager. Simply put, senior management must ''add value'' to the project management process.

Typically, many projects and project managers face issues that are beyond their effective control or delegation to resolve. For example, a union may raise issues with the job restructuring associated with the project; a vendor may not deliver an essential piece of software; a senior manager may not be prepared to release a key client/user person to assist in the project.

The senior management group associated with the project must be presented with these issues, alternative solutions, and clear statements of the impact on the project's deadlines, resources, priorities, quality, and so on for each of the alternatives to enable them to assist in resolving the problem.

To summarize, senior management must be placed in the role of problem solving as well as in the traditional roles of approval, monitoring, and review. To enable this, the focus for senior management must be on a critical set of information which is nontechnical and business-oriented.

Rapid Application Planning Sessions

An additional element of the team-based planning process is the use of highly structured project planning sessions. Just as the system development process is being shortened through the use of Joint Requirements Planning (JRP), Joint Application Design (JAD), and Rapid Application Development (RAD)—see Martin [1991]—which all entail the use of intensive team-based analysis, design, and development sessions, Thomsett Associates' approach to project planning and management involves the use of similar techniques.

The RAP sessions are convened by the project manager and involve any known team members, client area business experts, representatives of interrelated projects and key technical support groups (Computer Operations, Network, Data Base Administration). They follow the project planning process detailed later in this handbook and are facilitated either by the project manager or by an independent facilitator. Typical RAP sessions would last from one to three days.

Just as JRP and JAD sessions are supported by the use of CASE tools, the RAP sessions are supplemented by the use of automated estimation and scheduling software.

Apart from the advantages gained from the team-based approach, RAP sessions enable project managers to drastically shorten the traditional process of project planning. For example, to completely plan a $500,000, nine-month project requiring five development people for a Thomsett Associates client took one elapsed day.

The Critical Project Management Information Set

Throughout this handbook the emphasis will be on the integration of information required by all people involved in the project to understand the dynamic of the project.

As outlined in Figure 1.1, two sets of information are required to understand any project. The first if the technical information set. Depending on the nature of the project, this information describes the inherent technical content, issues, and development process associated with the project's deliverables or outputs. For example, in a computer software project, this technical information would include a data dictionary, data models, function diagrams, design charts, program specifications, test plans, training material, and so on.

Equally important to understanding a project—in fact, it is probably more important—is the project management information set. The project management information set is the framework for *all* key management decisions regarding the project. It is the link between the project team, project manager, project clients/users, and senior management. It is important to note that the content, structure and emphasis of the project management information set should be as *independent* as possible from the specific technical nature of the project.

The detailed contents of this set and the various procedures for developing and maintaining it are described in this handbook. However, a general overview of this information is:

- Project objectives.
- Project scope.
- Key external areas/groups/projects.
- Critical deadlines.
- Project returns/benefits.
- Project costs/estimates.
- Project risk.
- Project/product quality.
- Project development strategy.
- Project people.
- Project schedule/tasks.

As described in this handbook, the project management information set should be treated as the basis for a project contract/agreement/charter and as the major vehicle for change monitoring and control.

Only senior management can approve and alter the management issues related to a project.

Real-Time Project Management

Traditional project management was based on the interrelated concepts of a total/overall project plan for the entire project and a more detailed plan for the phase or subproject about to be commenced. Upon completion of the subproject or phase, the project management/planning cycle was repeated for the next phase or subproject.

While the simplicity of this approach is intellectually appealing, it is inappropriate for the more turbulent and unpredictable environment of the 1900s.

Contemporary project management must operate in real time using a micro- as well as a macroplanning approach. Real-time planning or microplanning involves the detailed planning of a project for a limited period, such as three to six months. Depending on the nature of the project (see Risk Assessment, Chapter 8) and the turbulence of the project environment, microplanning recognizes that many projects cannot be accurately planned over the entire development cycle because they depend on decisions that cannot be made until certain development work has been completed. For example, until detailed analysis is complete, the choice of appropriate design and technology strategies often cannot be completed. As a result, the detailed planning of the design and implementation phases of the project cannot be undertaken until the selection of the design and technology strategies.

The microplanning approach, coupled with the team-oriented RAP sessions, involves regular intensive planning sessions (at a minimum on a weekly basis), which develop detailed project plans for the next period (as distinct from the next phase or subproject), update the project management information set, develop problem solutions, and inform senior management of any potential variations for management review and approval. It should be noted that the frequency of the real-time planning process is heavily dependent on the risk of the project (see Risk Assessment, Chapter 8).

What is essential is that the project information set and the project plan *always* reflect the reality of the project. The more turbulent the project and its environment, the more "real-time" the project management process should be.

Whenever any component of the project information set is modified—for example, the risk changes or objectives are altered—the project manager and project team must stop the project and conduct a RAP session to evaluate the impact of the change, determine alternative solutions and options, develop alternative plans, and, if required, seek senior management and client approval to continue the project.

The Structure of the Handbook

These project management concepts underlie the approach to project management described in the remaining chapters of this handbook.

Chapter 2 introduces the basic project management model and the processes of project initiation.

Chapter 3 covers the essential project planning processes and the numerous issues such as estimation, Risk Assessment, and strategy included in Thomsett Associates' approach to project management.

The tracking and review of projects against plans are covered in Chapter 4.

Chapter 5 covers the issues of external groups and Project Agreements.

Chapter 6 provides some guidelines on Steering Committees and their roles.

Chapter 7 overviews the critical issues of return on investment analysis.

Chapter 8 contains a detailed Risk Assessment model for use in project planning.

Chapter 9 discusses some hints and advanced issues in quality and estimation.

Chapter 10 provides additional guidelines on project sizing and the special concerns of large and super-large projects.

Chapter 11 examines the project management impact of object-oriented development and the Rapid Application Development process.

Finally there are three Appendixes. Appendix A is a tutorial on Function Point metrics and estimation, and Appendix B provides a comprehensive model for determining and measuring quality requirements for a project. Appendix C provides an annotated list of the essential project management texts and journals.

References

W. G. Bennis, "A funny thing happened on the way to the future," *American Psychologist,* Vol. 25, No. 7, 1970.

P. F. Drucker, *Innovation and Entrepreneurship.* New York: Harper & Row, 1985.

F. E. Emery & E. L. Trist, "The Casual Nature of Organizational Environments," *Human Relations,* Vol. 18, 1965.

C. Handy, *The Age of Unreason.* Boston: Harvard Business School Press, 1989.

J. Martin, *Rapid Application Development.* New York, N.Y.: Macmillan, 1991.

T. Peters, *Thriving on Chaos.* London: Macmillan, 1988.

J. P. Womack, D. T. Jones, & D. Roos, *The Machine That Changed the World.* New York: Rawson Assoc., 1990.

2 Project Initiation

The first step of any journey is the most important.

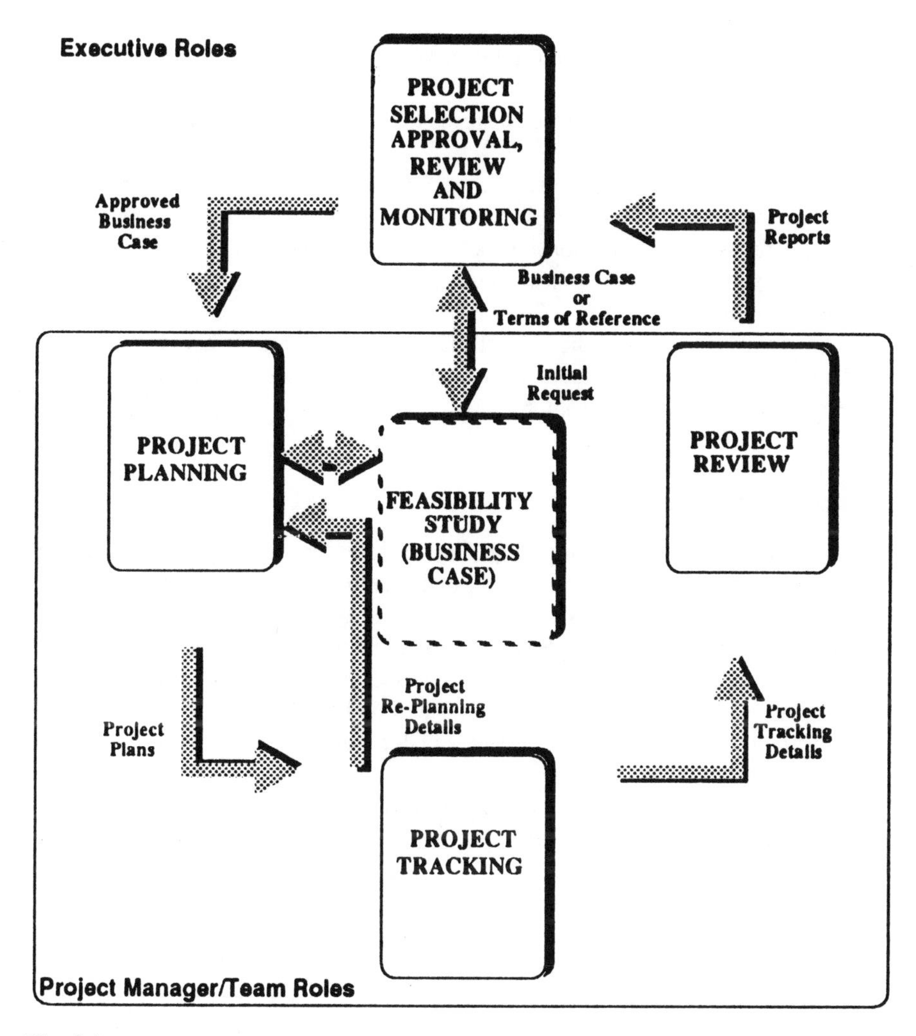

Fig. 2.1
Model of Project Management

The model of project management developed by Thomsett Associates involves four management processes and one system development process, as shown in Figure 2.1.

The four management processes are:

- Project selection, approval, and review.
- Project planning.
- Project tracking or monitoring.
- Project review.

The system development process is:

- Develop Terms of Reference/Business Case or Feasibility Study.

There are detailed descriptions of each of the five activities in Chapters 3, 4, and 5. However, in this chapter, we will explore the overall project management model, focus on the relationship between senior management and the project manager/team, and look at how the Business Case is produced during the Feasibility Study.

As a general rule, the project management approach involves a two-step project initiation approach. As in all journeys, the first step is the most important. (Step 2 is discussed in Chapter 3.)

Step 1: Project Initiation

This process involves a complex interrelationship between senior management, strategic planners, project managers, and systems or business analysts.

In approaching this step, it is important to understand that, while senior mangers are the critical people in the process of project selection, approval, and review, they depend on a set of critical project information. This information is developed primarily by the project manager and the business analysts to determine and confirm whether the project is viable from both a business/management and a technical perspective.

The way in which this would normally occur is based on the assumption that, in a strategic planning exercise, senior management would have already identified major projects that reflected their preferred corporate mission and direction. This strategic plan would also accommodate any new projects or initiatives resulting from external events after the completion of the strategic planning exercise (see Figure 2.2).

For most projects, the senior management would generally approve the first stage of the project initiation process—the completion of a Feasibility Study or Terms of Reference Study. This involves a high-level study of the project, including initial systems analysis with the aim of producing two key sets of information:

- The technical overview of the project with proposed solutions.
- The project's Business Case or Terms of Reference.

The development of the technical overview is not covered in this handbook. Suffice it to say, that, for a computing software project, the technical overview

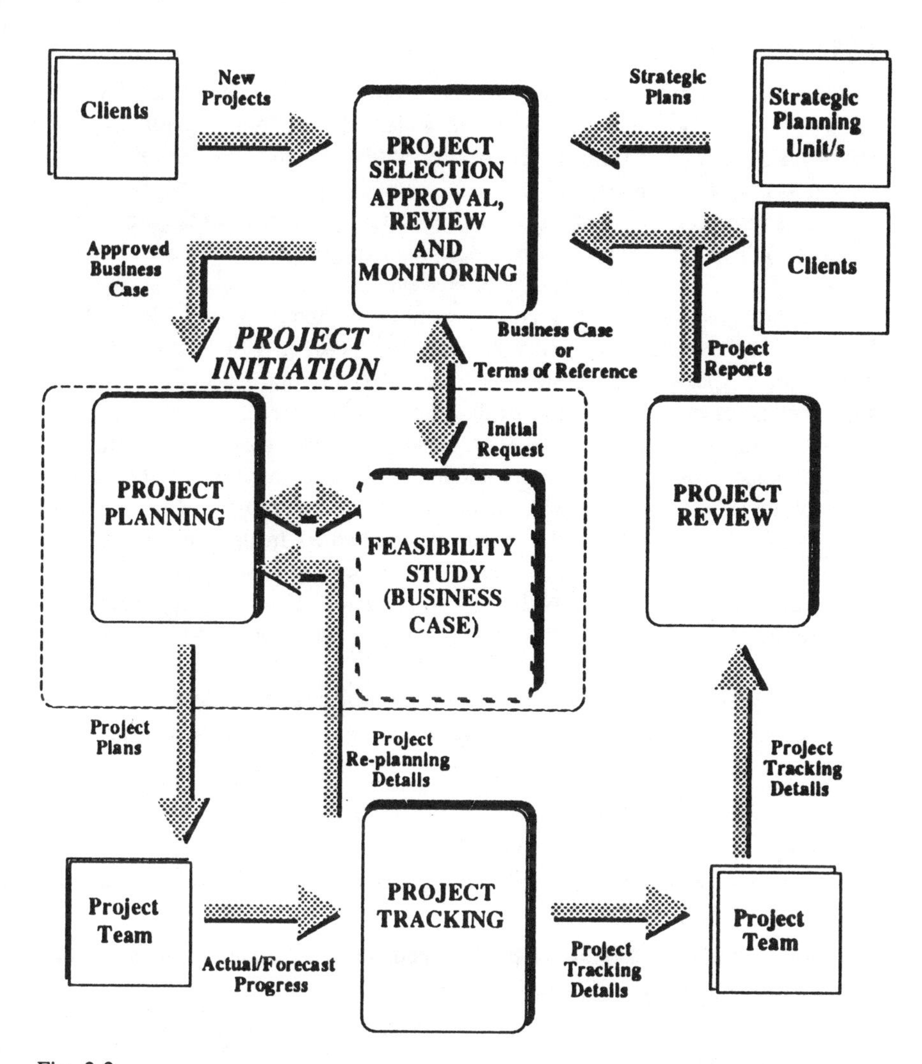

Fig. 2.2
Detailed Project Management Model

would include initial high-level data models, data flow diagrams, problem descriptions, and proposed alternative system designs or solutions. However, the quality of the technical overview is critical to the Business Case because it is the basis of estimation and quality requirements.

Business Case

The Business Case is the critical project management information set referred to in Chapter 1. Also known as Terms of Reference or Project Summary, it is produced as an output from the Feasibility Study, which is the first phase of a typical system development life cycle or methodology.

It is produced as a result of the initial planning sessions for the project (see Chapter 3). The typical process for developing a project Business Case is the production of an initial "draft" via the project planning process at the beginning of the Feasibility Study phase and its refinement during the progress of the study. A final version would be completed at the end of the Feasibility Study and submitted to the project sponsor and/or Steering Committee for approval to proceed to the second stage of project initiation—full system development.

The Business Case provides the "benchmark" or "baseline" for monitoring the project from initiation to completion. As already mentioned, it is normal for a project to be subjected to variations and change. The project's Business Case is the main *management* vehicle for monitoring and controlling the changes.

Ideally, the content and format of the project Business Case should be consistent across all projects within an organization. However, irrespective of the format, all Business Cases should contain the following set of information:

- *Project overview:* A brief description of the project and background.
- *Project scope:* A clear statement of the area/s of impact and boundaries of the project.
- *Project objectives:* A precise description of the project's objectives at the Strategic, Business, and System level.
- *Project returns:* The benefits that the organization can expect to obtain from undertaking the project.
- *Project costs:* The costs of the project (people, time, equipment, and so on) estimated over the development and operational cycle.
- *Project development strategy:* The overall partitioning and sequencing of the project into releases and subprojects.
- *Project risk assessment:* A formal assessment of potential risks associated with the project.
- *Project development schedule:* A series of project schedules and plans.
- *Relevant legislation:* A description of any relevant legislation or government policy associated with the project.
- *Stakeholders/key groups:* Key groups and organizations outside the project manager's direct control upon which the project is dependent.
- *Project staffing:* Clear statements of the assumptions made regarding suitable project people.

- *Interrelated projects:* An overview of any other projects dependent on or interrelated with the proposed project.
- *Assumptions and constraints:* Any assumptions or constraints, such a deadlines or technology choices.
- *Project quality:* A statement including measures of the required quality of the product.

Clearly, the production of this information requires not only skilled people, but also a relationship between the project manager who is generally responsible for the estimates, development strategy, risks, and costs and the business or systems analyst who would undertake the detailed modeling of the project's scope, objectives, returns, and dependent projects. Ideally, the Feasibility Study is a joint development effort involving both business and technical experts. As discussed throughout this handbook, the involvement of key groups impacted by the project and specialist groups—such as Finance, Administration, and Human Resource—would further enhance the quality of the study and "buy-in" by the major project participants.

As shown in Figure 2.3, upon receipt of the project's Business Case and Feasibility Study, senior managers decide whether to proceed to Step 2 of the project planning process, Detailed Project Planning and subsequent development, or to cancel any further development of the project.

A Note on Initial Estimates

It is problematical that projects are justified at the worst time—at the beginning. Cost-benefit analyses and other information used to approve projects are, in general, only "ballpark" figures. As discussed throughout this handbook, as the project proceeds, more accurate estimates will become available. The Business Case must be constantly updated and monitored to reflect the current reality of the project. Any major variation of the figures, such costs and benefits used to initially justify the project, should be resubmitted to the Project Sponsor and/or Steering Committee (see Chapters 4, 5, and 6).

An initial planning session using the processes described in the following chapter is conducted at the commencement of the Feasibility Study phase and focuses on developing estimates, schedules, and so on for the Feasibility Study Phase. It also involves planning the overall project at a high level using "ballpark" estimates.

It is essential to understand that the approval of a project is a business rather than a technical decision.

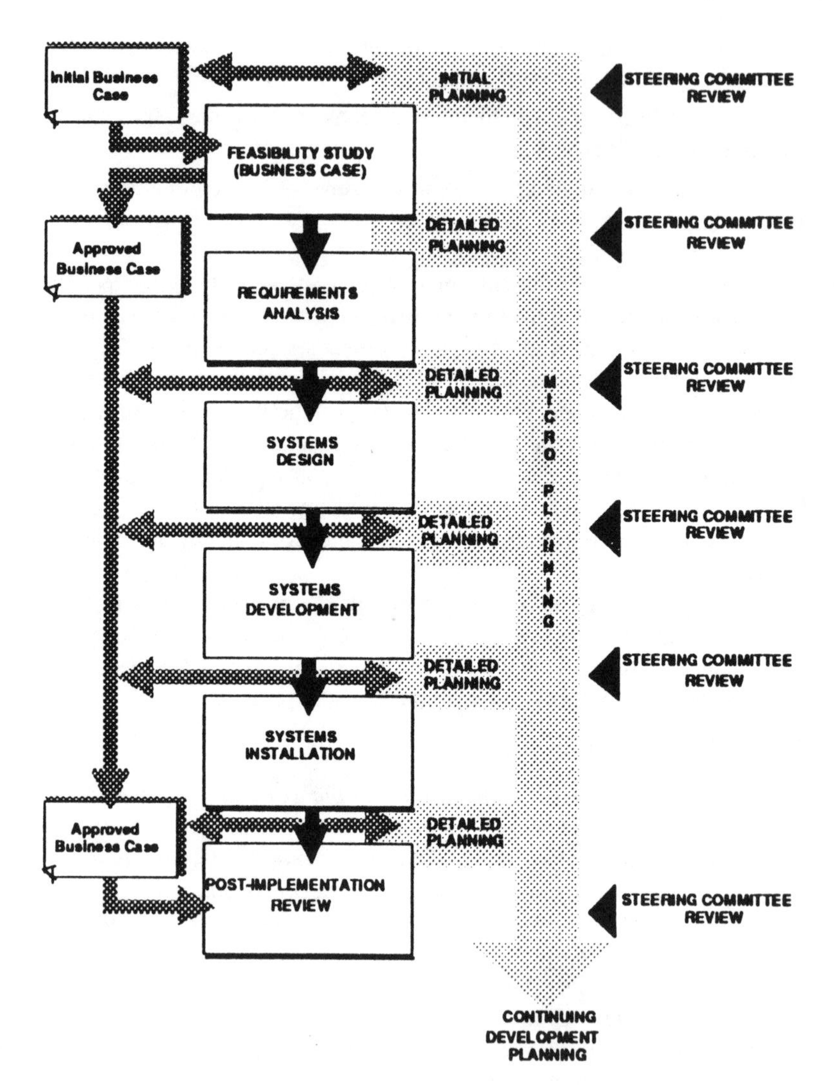

Fig. 2.3
Project Initiation and Development

3 Project Planning

Planning is working—working is not planning.

At the end of the Feasibility Study and prior to the commencement of the full development cycle, the project manager and team should undertake the second step of the project planning process—a detailed project planning session.

As discussed in Chapter 1, the preferred approach to this planning process consists of team-based, Rapid Application Planning sessions. Various techniques, which ensure that these planning sessions are conducted smoothly, are contained in Chapter 9. However, the key concept is the use of a systematic six-task planning model detailed below.

As shown in Figure 3.1, project planning incorporates the tasks of:

- Reviewing the Business Case.
- Selecting an appropriate project development strategy.
- Conducting a formal Risk Assessment.
- Selecting the project's tasks.
- Estimation of the tasks, and
- Scheduling of tasks and people.

The resultant project plan is then the basis of both tracking and reporting. Actual work/process on tasks is compared to planned progress and major deviations and resolved. In many cases, this may involve a replanning process. Project reporting gathers information across all tasks within the projects for review by users, other related project managers, and corporate management.

Project planning is performed at the beginning of each project and repeated at regular sessions during the project via microplanning and RAP sessions. This is the Real-Time Project Management concept discussed in Chapter 1.

Depending on the size of the project, the project manager or business analyst produces an initial Project Plan for the project while completing the Business Case. However, detailed planning and scheduling do not normally commence until the beginning of Requirements Analysis or the first phase of the product development cycle.

The project planning approach recommended is the real-time planning approach introduced in Chapter 1, wherein the detailed planning and production of a project plan or schedule is undertaken at the beginning of each phase and applies only to that phase. The overall project plan provides broad details and estimates for all phases; however, in many projects the overall project plan should be considered as a rough approximation. Simply put, until development alternatives are evaluated, the project manager cannot accurately plan System Design and subsequent phases.

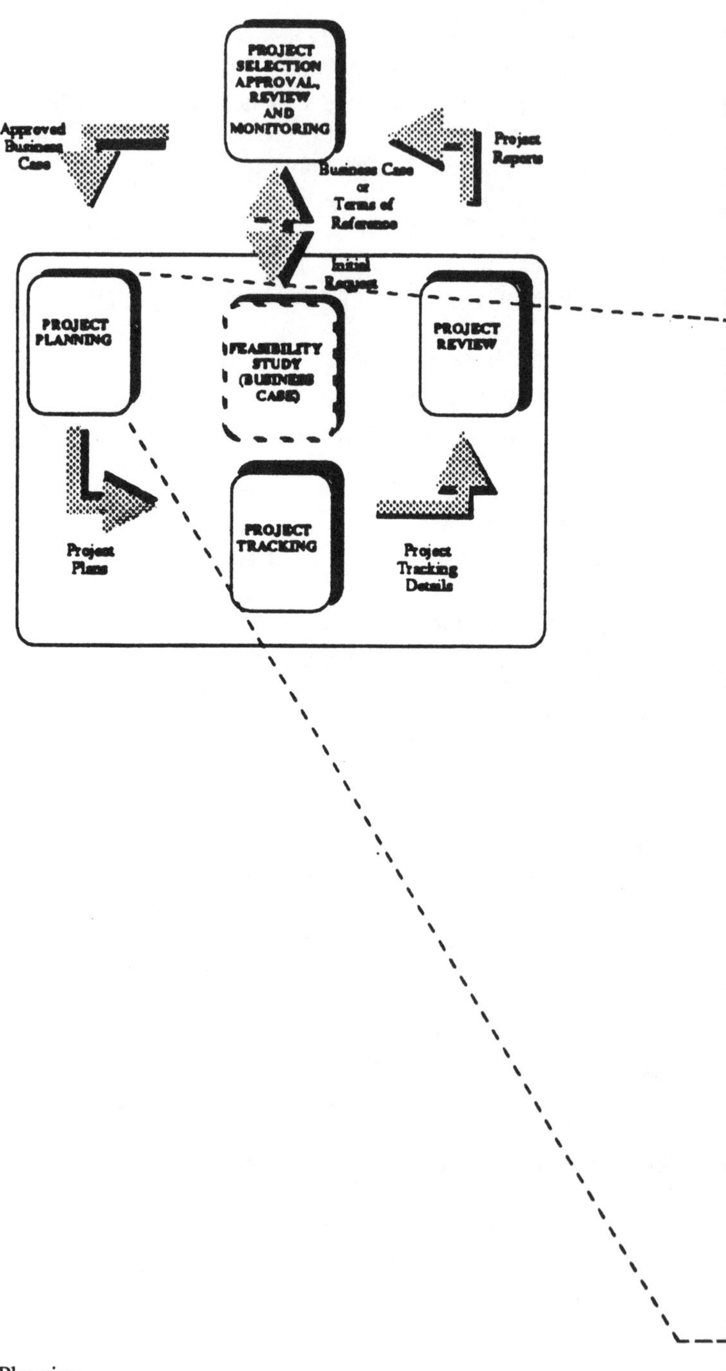

Fig. 3.1
Detailed Project Planning

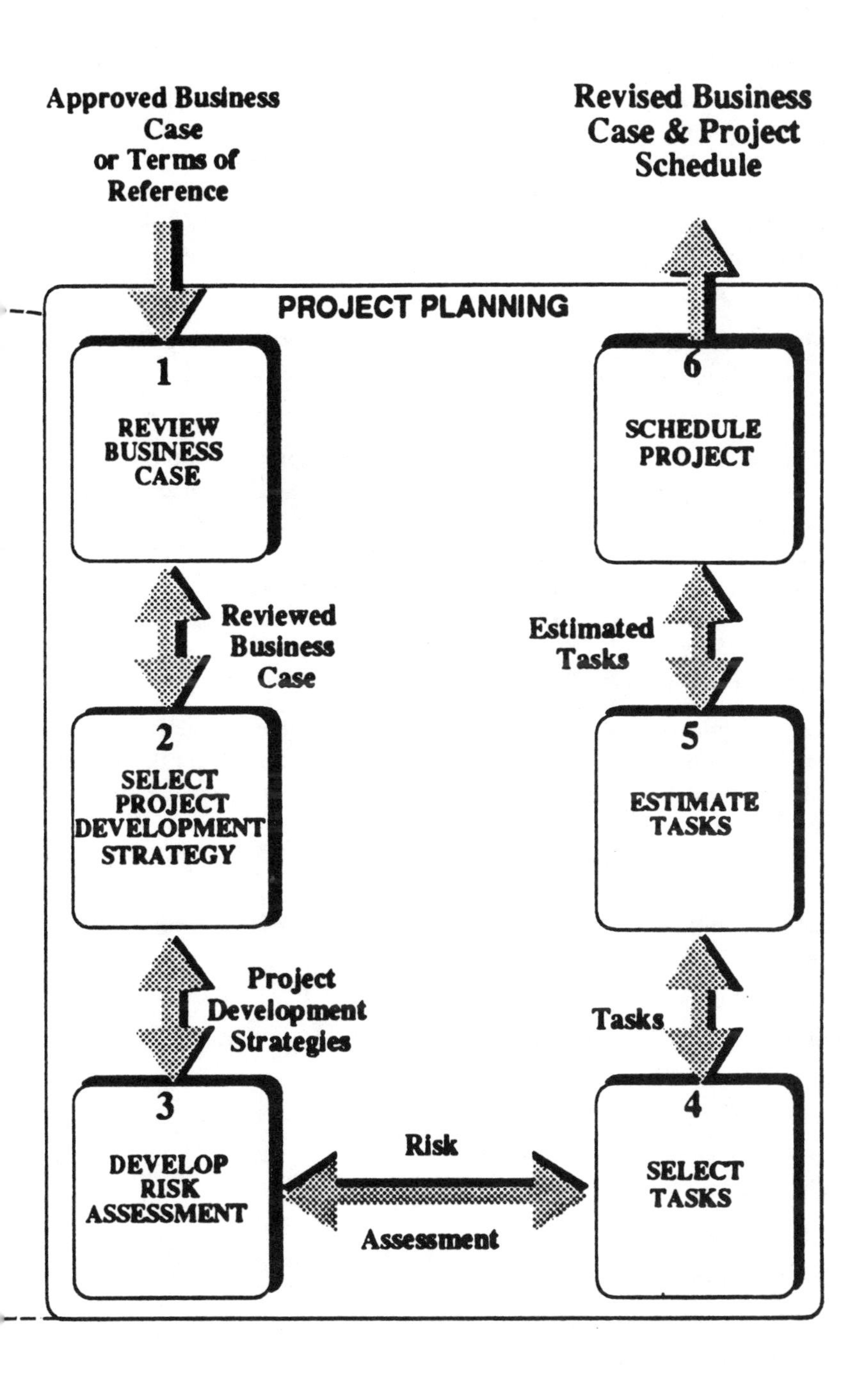
Approved Business Case or Terms of Reference
Revised Business Case & Project Schedule
PROJECT PLANNING
1
REVIEW BUSINESS CASE
Reviewed Business Case
2
SELECT PROJECT DEVELOPMENT STRATEGY
Project Development Strategies
3
DEVELOP RISK ASSESSMENT
Risk
Assessment
4
SELECT TASKS
Tasks
5
ESTIMATE TASKS
Estimated Tasks
6
SCHEDULE PROJECT

Step 2: Detailed Project Planning

As outlined in Figure 3.1, detailed project planning has six interrelated and iterative tasks. While these tasks are depicted serially, they are highly iterative and will often be done many times before a suitable Project Plan can be produced. For example, the project team may have made some assumptions about a new member's skills during estimation of some tasks. Upon scheduling the person to the tasks, the actual person may have different skills, which will require a revision of the estimates.

Project planning is a team-driven process; all team members should be active in planning their project.

As outlined in the Introduction, Thomsett Associates' approach to project planning is based on a participative planning session involving the project manager, his or her team, representatives of key external groups and projects (see Chapter 5), and any technical support groups. If the project manager does not have a team allocated during the initial planning phases, he or she should have at least one peer and any identified key stakeholders involved in the RAP session.

Task 1: Review Business Case

The first process in detailed project planning is to review and if necessary retrofit the Business Case should a Business Case not be available for the project.

The Business Case should have been produced during Step 1, Projection Initiation (Feasibility Study) in the project development cycle.

As described in Chapter 2, it should contain:

- Project Overview.
- Statement of Project Scope.
- Statement of Project Objectives.
- Initial Returns/Benefits Analysis.
- Initial Project Cost Estimates.
- Project Development Strategy.
- Initial Risk Assessment.
- Proposed Project Schedule.
- Any Relevant Legislation.
- Stakeholders/Key Groups.
- Dependent or Interrelated Projects.
- Assumptions on Project Staffing.

- Assumptions and Constraints.
- Project Quality.

If the project being planned does not have a Business Case, then the project planning process must include the gathering of this information (in particular, the risk assessment, objectives, scope, and high-level technical models.) For smaller projects, the retrofitting of the Business Case may be done during the initial planning session. For large projects, a full Feasibility Study may be required. This will involve business analysts, systems analysts, and clients.

Task 2: Select Project Development Strategy

A distinction must be made between development methodology or tasks and a project development strategy. A project development *strategy* governs how the overall system development process is undertaken, while the *methodology* is the specific tasks to be undertaken during the development cycle. The appropriate strategy is highly dependent on the risk and the size of the project.

Based on the work of Paul Melichar [1982], there are four basic system development strategies:

- *Phased or Monolithic:* This involves developing the system as a whole with each phase as a stand-alone activity, and subsequent phases are not commenced until preceding phases are complete. For instance, design is not commenced until analysis is complete. This strategy is suited for low-risk projects of less than six months in duration and with small teams (fewer than five people).
- *Release or Version:* This involves breaking the system into semi-independent subsystems and producing an entire operational subsystem or product as Version #1 or Release #1, then a second subsystem added as Version #2 or Release #2, etc. The user will be given "semioperational" components of the system. For example, Version #1 could be all the input components (including editing, etc.) that create live data files. While Version #2, the output components, is being developed, the users can access the live data via query languages. In Version #3, the full data base facility could be added. This is termed *Sequential Release*. Alternatively, the project team is broken into subteams, each of which produces a subsystem in a parallel or concurrent manner. This is termed *Concurrent Release*. Each release uses the methodology as per the Phased Strategy.
- *Fast Track:* This involves producing a *production* "prototype" of the system as fast as possible. This can be achieved by minimizing adherence to standards and/or by use of application generators or high-level languages. The first operational "prototype" is redeveloped through a series of rewrites. Clearly, this is a high-risk strategy and requires extensive negotiations with management and project stakeholders.

- *Hybrid:* This is a variation of Concurrent Release. The hybrid strategy really involves a series of releases or subprojects with each subproject or release using a differing development strategy. For example, a high-risk subproject may be developed using the Fast Track strategy while a low-risk subproject is developed concurrently using the Phased Strategy.

The following are some guidelines for selecting an appropriate project development strategy:

Project Size	*Risk Assessment*	*Strategy*
Less than 3 months (duration)	Low	Phased
	Medium	Release
	High	Fast Track
3–6 months	Low	Phaseed or Release
	Medium	Release
	High	Release or Fast Track
Over 6 months	Low	Release
	Medium	Release
	High	Hybrid or Fast Track

Generally, with a team size of greater than 5–7 people, the project has to use one of the variations of the Release or Hybrid strategies. Simply, as shown by Fred Brooks in his classic book *The Mythical Man Month* [1975], it is more efficient to add additional people to subdivided projects/tasks.

Figure 3.2 depicts the differences between these strategies and how the project schedule appears depending on which project development strategy is used.

It is essential to recognize that, for many projects, the strategy may change at various stages of the development process. For example, the project may have a high-risk Requirements Analysis phase requiring the use of the Fast Track Strategy (which may involve the developing of prototype screens etc.). However, once the client's requirements are determined, the project may move to a low-risk Design, Development, and Implementation stage. For the remainder of the project a more conservative strategy such as Sequential Release or Monolithic may be implemented.

The selection or alteration of the project development strategy must be negotiated with senior management and clients.

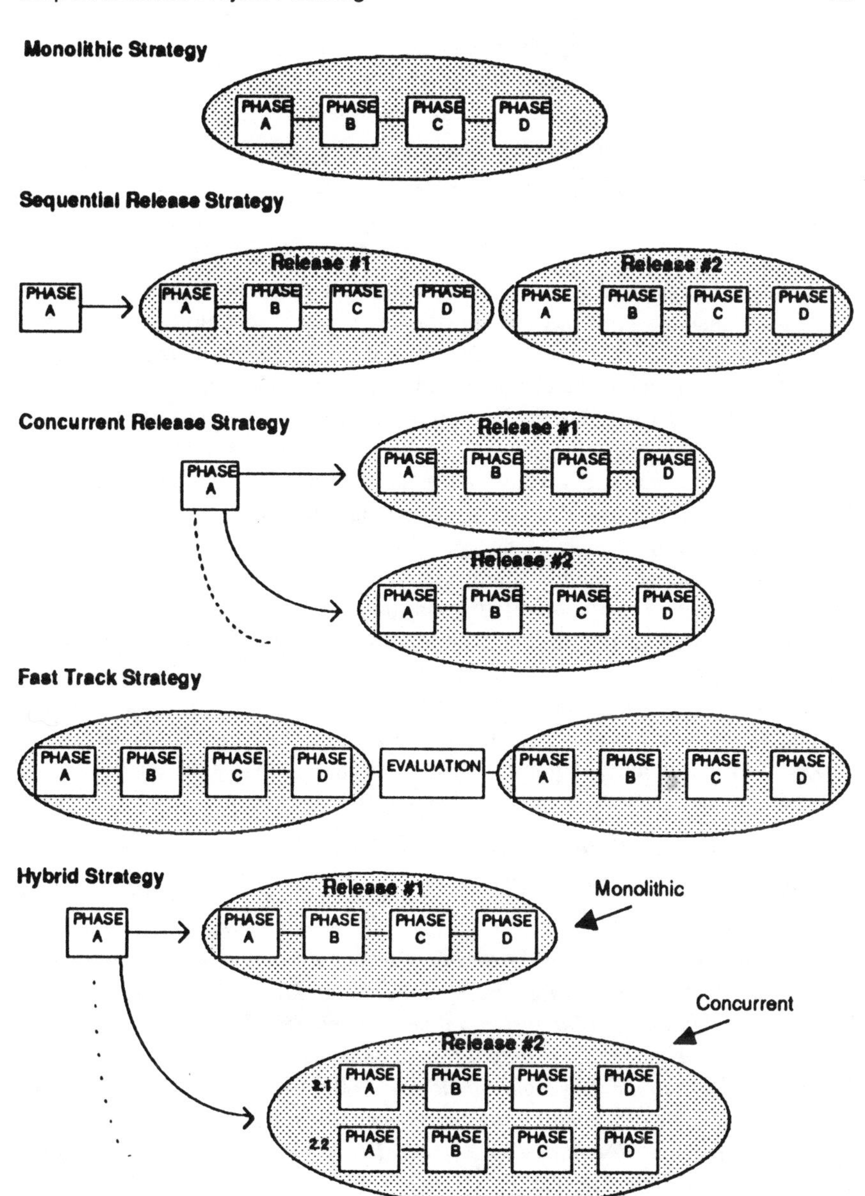

Fig. 3.2
Project Development Strategies

Should the Release or Hybrid Strategy be chosen, the system can be "packaged" into Versions at three primary points:

- At the end of the Business Case Study.
- At the end of detailed Requirements Analysis.
- At the end of Systems Design.

It should be emphasized that, prior to partitioning the system into subsystems, the project manager and team should develop a representation of the system (e.g., data flow, flow chart, or structure chart) sufficient to ensure that the releases have minimum data interdependence. The earlier the system is partitioned, the higher the risk that the subsystems may not be partitioned cleanly. This is to ensure that each release can be treated as a virtually independent subsystem. If the organization has a Data Administrator or Information Resource Manager, that person would be involved as a key stakeholder in the partitioning of the system during project planning.

In addition, with multiple releases of a system being developed concurrently, it is also very important to monitor changes to releases while under development. Since it is data that generally links subsystems, it is essential that the data common between subsystems is managed on behalf of all subsystems to ensure that alterations in common data requirements from one subsystem is understood by all other subsystems. The formal use of Change Control (refer to Chapter 4) is particularly important with this strategy. Similar issues arise should common business functions be shared across the subsystems.

A Note on Rapid Application Development (RAD)

With the advent of CASE software, which assists in the storing and maintenance of key project deliverables such as data models, data flow diagrams, structured design diagrams, file designs, program specifications, and so on, the project developments strategies of Release and Fast Tracking can be supported by CASE. The partitioning of a system into small releases, which are then developed using Sequential or Concurrent Release strategies, has been termed Rapid Application Development by some vendors (see Martin [op cit]). Some versions of this strategy suggest partitioning the system into releases that can be developed and implemented within 90–120 days. For all intents, RAD is a variation of the Sequential Release strategy.

Task 3: Develop Risk Assessment

During the project planning process, the risk assessment normally completed during the development of the Business Case is reviewed and updated by the project team. Risk assessment is a formalized subjective assessment of the probability of project success. Risk assessment has an obvious impact on the management style, team structure, use of methodology, strategies for system development, and, most importantly, the business decision to approve the project.

Simply, the greater the risk of the project, the higher the probability that estimates, schedules, and planning will be incorrect and that the project will move "out-of-control."

The risk of a project can be established by considering the following criteria:

- System or product complexity.
- Client or target environment.
- Team environment.

A full checklist of items that will have an impact on each of these areas is contained in Chapter 8.

System Complexity

To evaluate the complexity and risk of a software project, we borrow from an IBM technique called Function Points (see Appendix A) and consider its data complexity, that is, how many imputs, outputs, inquiries, logical internal files, and shared files are involved in the system.

Other aspects affecting system complexity or risk are:

- Function and algorithms.
- Complex control, decision exception, and/or mathematical operations.
- Existing client procedures and impact.
- Significant impact on jobs.
- Performance requirements.
- High data volume, fast response time to minimize network use, CPU, etc.
- Technology requirements.
- Substantial use of tailored or special hardware/software.

Target Environment

The complexity or risk of the user environment is related to the following areas:

- The number of user sites (sections and departments, agencies, branches, and installations) involved in developing and implementing the system.
- The level of user knowledge and participation in the application and the project development process.
- The priority and impact of the application within the user area.

• The need for physical restructuring of offices, development of new sites, etc.

Team Environment

The complexity or risk of the team environment is related to the following areas:

• Schedules, whether fixed or flexible.
• The experience and likely stability of the project team.
• The development and estimated time frame of the project.
• Use of outside vendors/contractors.
• The physical team/project environment.

It is mandatory that, throughout the system development process and especially during project planning, the project manager consider these project risk criteria using a formal questionnaire as contained in Chapter 8.

If the project manager considers the combination of any of these factors is significant and contributes to the degree of risk of the project, he or she is encouraged to consider the following actions:

• Take steps to limit the scope of the project to reduce its complexity.
• Document the areas of complexity in the Project Plan and allow for additional time/resources.
• Raise a formal Risk Memorandum (see Chapter 8) that details the high-level factors, identifies their possible impact and actions/options available to reduce that impact or reduce the risk factor.

As discussed in Chapter 8, it is imperative that the management of project risk is seen as a proactive process. For example, prior to the commencement of the full development cycle, the project manager should negotiate with the Steering Committee, key stakeholders and sponsor to minimize the high-risk factors.

The management and minimization of project risk is the responsibility of all involved parties in a project.

A Note on Software Metrics

The risk-assessment process introduced in this handbook is primarily a subjective one. The process of quantifying the various risk factors is termed Software Metrics. For example, subjective experience indicates that there is a substantial difference between an inexperienced and experienced programmer. Therefore, using a subjective model, a project with inexperienced programmers would entail a higher risk than one using experienced programmers (everything else being equal). Attempts to quantify the difference between experienced and inexperienced programmers have lead to inconclusive results. For example, Sackman et al. [1968] measured differences of 26:1; DeMarco and Lister [1987] put the difference as 10:1 and Jones [1986] placed the variance at 2:1. Clearly, there is a problem.

It is given that over 100 factors have been determined as significant in affecting project productivity and risk, and that many of these factors are, for all practical purposes, unmeasurable (for example what is the quantified impact of office politics between two key client groups?). As a result, the area of software metrics remains an imperfect one.

Perhaps the most effective use of software metrics is to use the available measures (from as many sources as possible) to support your subjective risk assessment and to adjust the estimates where possible. Software metrics are discussed further in Chapter 8.

Task 4: Select Tasks

Within a typical system or product development phases are broken into tasks and tasks are broken into sub/tasks (see Figure 3.3). This is often termed a *work breakdown structure* or *methodology* or, more simply, a *task list.*

While the process of identifying the tasks required for a project is conceptually simple, it is a vital step in the process since it provides the basis for bottom-up estimation and developing a detailed project schedule.

Most organizations have developed a standard work breakdown structure for different classes of projects such as in-house development and package selection and implementation. Depending on the risk and nature of the project, the project

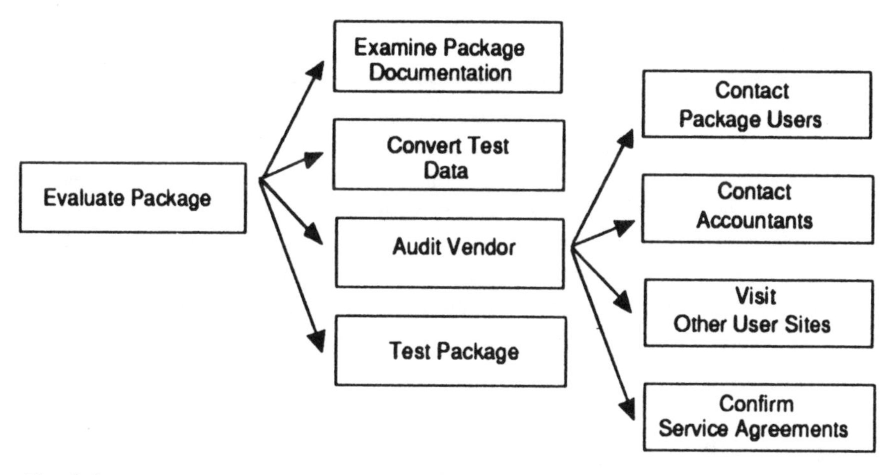

Fig. 3.3
Work Breakdown

manager is encouraged to combine, delete, or add tasks or subtasks detailed in the methodology. For example, additional tasks may be required for selecting hardware while the data base subtasks may not be required. This tailoring will be done by the project manager assisted by the relevant specialists in the organization.

Should the organization not have a standard methodology, or if the project is not suited to the standard methodology, the planning team should brainstorm the task list. It is important in all cases not to confuse selecting the tasks with scheduling tasks. The tasks should be selected in any order. The sequencing (scheduling) of the tasks is done later in the planning session.

Ideally, no project activity phase (task, subtask) should have an elapsed duration of more than 10–20 calendar days. This is not to be confused with work effort. For example, a task could legitimately require two people for 15 elapsed days, that is, it involves 30 work effort days. When a task would exceed 20 elapsed days it should be broken down into smaller tasks.

It is recognized that it may not always be possible to strictly observe the "10–20-day" rule. The main aim of tasks is, again, to break the phases into manageable, measurable portions.

The process of work breakdown or task listing is essential for estimation because it ensures that the team understands what work has to be done. One of the most common causes of poor estimation is simple failure to list all tasks required. In the RAP session, the use of experts and full participation by the team and stakeholders usually results in an accurate task list.

Task 5: Estimate Tasks

A number of alternative estimation techniques are available to the contemporary project manager. The most widely used is the *macrotechnique* of Function Points (see Appendix A). This technique can derive an estimate for the total development effort and the microtechnique of bottom-up or task-based estimates, which involves estimating tasks and "rolling-up" the tasks into a phase total. The *microtechnique* is required for developing project schedules.

As shown in Figure 3.4, the typical approach is for an initial macrotechnique estimate to be developed using Function Point, and this is then used to correlate a second estimate developed bottom-up via a microtechnique such as Wide-Band Delphi (see Chapter 9). The task estimates are aggregated into a phase estimate which, if the organization has a standard methodology, can be used to derive another total development effort using the phase percentages as a guide.

Estimating Actual Effort

For each task/subtask in each phase of the project, the project manager and team must produce an estimate of work effort and elapsed days required to complete the task/subtask.

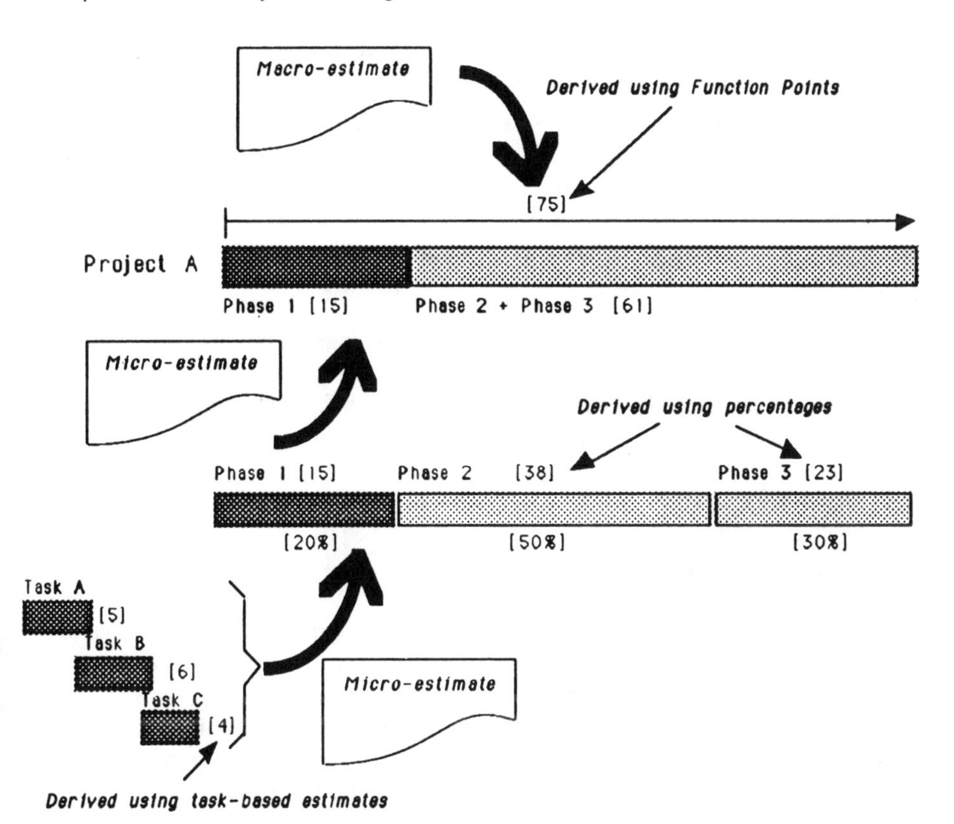

Fig. 3.4
Macro- and Microtechnique Estimation

Until projects within an organization begin to conform to a standard methodology (i.e., common task naming) and actual work effort and actual elapsed duration are tracked and recorded against a commonly named set of tasks, a formal estimating history will generally not be available. In the future, the project manager should be able to use output from the Project Tracking process (described later) from all projects using the methodology to obtain guidelines as to how long the standard phases/tasks took on similar projects, often called a metric data base.

Until this estimating history is available, four broad guidelines are suggested:

- The smaller and more specific the task/subtask activities, the easier they are to estimate.
- The greater the risk of the project, the greater the possibility of estimating error.

- Estimate by teams—for example, during the task definition team meeting—the project manager should seek estimates from team members and, as a rule of thumb, average all the estimates given for each task/subtask (see Wide-Band Delphi).
- Estimate assuming one person is doing the task, and estimate in work effort, not in duration (see later in this chapter). The mapping of work effort to elapsed duration and the allocation of actual resources is done during scheduling.

It is essential to state estimates as a range than a single figure, especially during the early phases of the project development cycle. The use of Sensitivity Analysis techniques is highly recommended. This technique involves making estimates ranged into three figures:

- Optimistic/best case.
- Realistic/likely.
- Pessimistic/worst case.

The *Optimistic* estimate is based on "everything" going better than expected. The *Realistic* estimate reflects the likely situation, and the *Pessimistic* estimate is based on the worst case scenario.

Then, based on Risk Assessment, the project manager should select one of these figures as a base for scheduling. For low-risk projects, the optimistic or the realistic estimates are the estimates used for scheduling. The higher the risk of the project, the more the realistic or pessimistic range would be used. However, all three sets of figures are included in the Project File (see Chapter 5).

In many cases, the Realistic estimate is the most useful estimate for scheduling; however, certain tasks may be of a higher risk than others and may require the pessimistic estimate. In other words, by doing an informal risk assessment of each task, the low-risk tasks are scheduled using the Optimistic estimate, medium-risk tasks use the realistic estimate, and high-risk tasks are scheduled on the pessimistic estimate.

An alternative method borrowed from the engineering area is to input all three estimates into the following equation:

$$\text{Expected} = [\text{Optimistic} + (4 \times \text{Realistic}) + \text{Pessimistic}]/6$$

This gives a single estimate (expected), which reflects the range and distribution of all three initial estimates.

Estimating Actual to Elapsed Effort

One of the major causes of poor estimation is involved in the difference between *work* or direct effort (effort hours) and the *elapsed* or calendar duration (elapsed hours) required to expend the effort. The process of estimation involves first estimating the work effort required, then estimating the elapsed or calendar duration involved to expend the work effort. For example, a process may be estimated at four hours work effort but, because of other commitments (see later) the four hours may be scheduled in two two-hour slots over two days. Therefore, the elapsed duration is two days. Since all project scheduling is in elapsed or calendar days, project managers must take into account this distinction between work effort (WE) and elapsed duration (ED).

Typical reasons for the difference between work effort and elapsed duration include:

- Re-work or defect repair on previously completed dependent tasks.
- Holidays, weekends and public holidays.
- Consulting or coaching other team members.
- Rostered days off.
- Non-project administration.
- Non-project education and team coaching.
- Non-project meetings.
- Interruptions including phone calls.
- Non-project paperwork.
- Wait time for meetings, results.
- Switch time, e.g., the time required to "switch" between tasks.

It is not unusual for these factors to account for 30–50 percent of an elapsed day. Therefore, it is acceptable to find an WE of one day taking an ED of two or three days depending on the particular project and administrative environment.

Further, the amount of "lost" time varies with the level of the person performing the task. For example, a 30 percent loss is typical for a junior programmer/analyst while a 50 percent or higher "loss" may be typical for a senior analyst or team leader who would spend more time in consulting, administration, coaching junior team members, and nonproject meetings.

Task 6: Schedule Project

This activity is carried out by the project manager, the team members, user representatives, and relevant specialists during the RAP session. It involves the allocation of resources to each task shown on the project's task list to produce a formal chronological Project Plan or Schedule.

This activity can be quite complex since there is typically a number of sequencing and resourcing options. For example, tasks may be sequenced linearly or concurrently depending on resource availability and task interrelationships. It is another example of how a team is able to develop more alternatives than a lone project manager.

When resources are to come from outside the project team, such as operations and client area staff, the project manager will of necessity have to liaise with other areas of the organization with regard to the people who will do the work, their availability, and whether the likely duration of the task is realistic. Ideally, these people will participate in the RAP session and will make an up-front commitment to release key resources while the project schedule is being developed in the planning session.

The method for carrying out this process is:

- Develop a first-cut network diagram showing which tasks must be complete before other tasks can commence (task dependency) and which tasks can be performed while other tasks are being completed (see Figure 3.5). This network is based on task/deliverable dependancies where one task requires the output or deliverable from another.
- From the task list take each task and, maintaing in chronological order of execution, allocate persons to it and pencil it on the draft schedule. The planning team should bear in mind that:
- Some tasks may require specialized skills or training.
- Tasks that depend on completion of prior tasks are not to be scheduled to precede or overlap those tasks.
- Allowances for on-going staff training, furlough, rostered days, sick leave, and overhead should be reviewed when allocating resources, if more than one person is working on the task. If necesssary, extend the likely duration to cover this. Avoid making too tight an assessment of likely duration to avoid having tasks consistently running over time; a reasonable amount of "insurance" should be built in.
- Adequate provision (about 10 percent of effort) should be made in the plan for Quality Assurance activities (see Chapter 9).

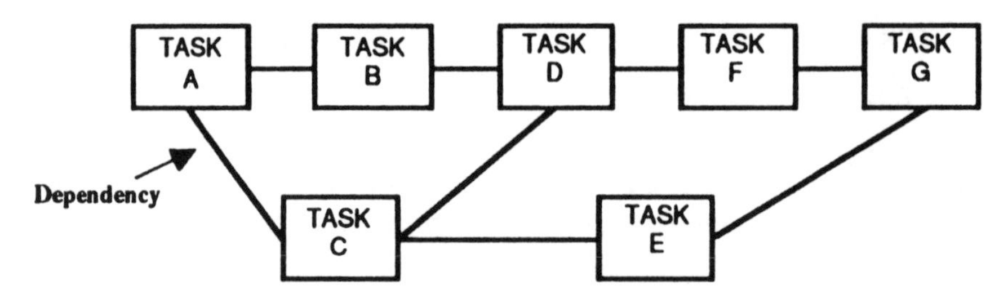

Fig. 3.5
First-Cut Network Diagram

- In some cases, assumptions regarding resources may have been made during estimating, such as the availability of senior analysts. Should the actual resource have different skills or availability, the estimates will need to be revised.
- When resources come from within the project team, allowance for planned or unplanned support of existing applications have to be made. Again, if necessary, extend the likely duration to cover this. For example, assigning a task of six full working days to an analyst who will spend half his/her time on other work, makes that task one of twelve working days (see Estimating Actual to Elapsed Duration).
- Caution is needed when assigning two concurrent tasks to the same person. If it is necessary to do so, extend the durations accordingly to allow for switch time.
- When the resource for a task comes from outside the project team, the likely duration may need to be an estimate, subject to later confirmation, and the person to whom the task will be assigned may not be known.
- The project manager should assign to himself/herself only tasks of a management nature or those that can be carried out without detriment to his/her management of the project.
- When all task estimates (in elapsed days) have been charted on the draft Schedule, it must be rechecked to ensure that all interdependencies have been allowed for and that, where tasks overlap and have been assigned the same resource, the assigned person will be able to carry out those tasks concurrently.

When the draft schedule is completed, a final eye-check should be made to ensure that tasks assigned to the same person do not overlap or, if they do, that it was intended to be that way and that the person will be capable of carrying out the tasks concurrently. This is called a Resource Dependency.

At this stage, the tasks, estimates in elapsed effort and resources should be input into a computer-based scheduling tool where the package will calculate the critical path for the project. The critical path is the longest path (in elapsed time) through the network. For example, in Figure 3.5, Task D is dependent on Task C and Task B. Since Task B takes four days and Task C is estimated to take three days, the earliest that Task D can start is four days after the completion of Task A. Task C has one day of float or slack and Task B is on the critical path. Where required, the project manager and team can reschedule or reallocate resources to minimize float activities and the schedule.

Scheduling software will also produce a Gantt chart (see Figure 3.6). The Gantt chart shows the tasks against the time line and indicates the tasks that are on the critical path and those that have float or slack.

All contemporary scheduling software packages also produce task lists by resources, resource loading, and other critical planning and tracking information. As discussed in Chapter 4, the project scheduling package can also be used as the basis for project tracking and reporting.

However, these tools are a classic example of "garbage-in/garbage-out." If the planning process is flawed, the resultant schedules and other information produced by these tools is simply false and dangerous.

A Note on Scheduling Software

A number of PC-based scheduling tools can be used to develop network, Gantt chart, and other useful project management documents. Most can also be used as a project cost and progress tracking tool as well. Thomsett & Associates suggests Microsoft Project for Windows 3.0 and Apple Mac's and MacProject II as tools that are easy to learn and use. These packages are powerful enough to provide most features required by project managers. For a comprehensive evaluation of other packages refer to Lois Zell's book, Managing Software Projects *[1990], and check your computer press since there are new tools being announced monthly.*

Real-Time Planning Revisited

The real-time approach to planning involves the repeating of the planning process a number of times during the project. In addition, should the project move

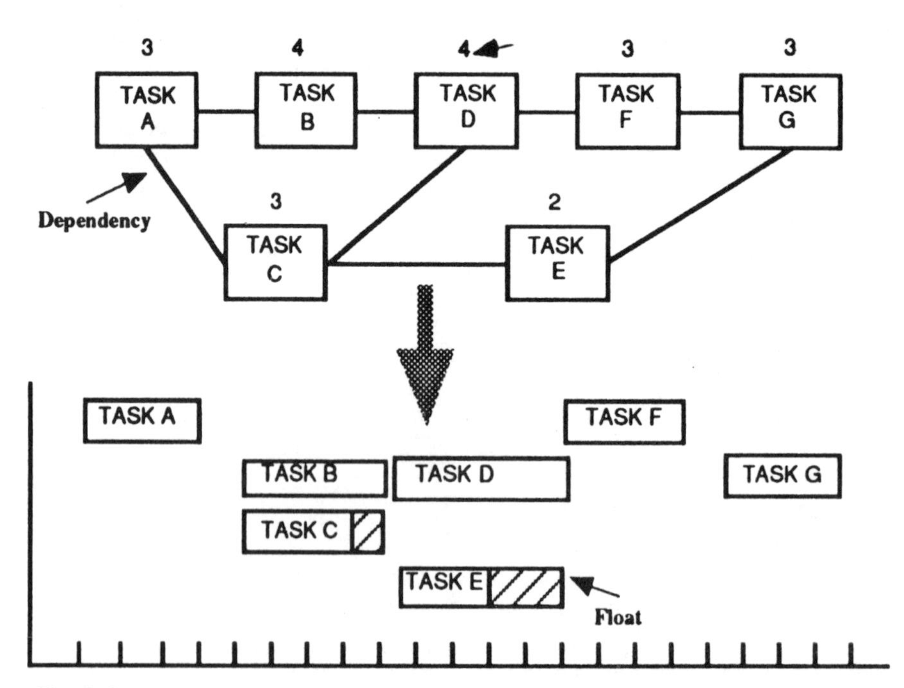

Fig. 3.6
Gantt Chart/Schedule

''out of control,'' as described in Chapter 4, the planning cycle will need to be invoked again.

It is very important to schedule only those phases/tasks that can reasonably be scheduled. While the use of computer-based scheduling packages can make a rescheduling task easier, there is little value in scheduling to the day a task that is, for example, dependent on 20 other tasks completing on time. In addition, the further away a task is in elapsed or calendar days, the less likely the date will be met.

Simply put, schedule at a detailed level for three to six months or to the end of the current phase in advance. The remaining tasks could be aggregated to phase level and scheduled in a detailed manner at a later planning session.

References

F. Brooks, *Mythical Man Month,* Reading, Mass.: Addison-Wesley, 1975.

T. DeMarco & T. Lister, *Peopleware*. New York: Dorset House, 1987.

C. Jones, *Programming Productivity*. New York: McGraw-Hill, 1986.

H. Sackman, W. J. Erikson, and E. E. Grant, ''Exploratory experimental studies comparing online and offline programming performance,'' *Communications of the A.C.M.*, Vol. 11, No. 1, January 1968, pp. 3–11.

L. Zells, *Managing Software Projects*. Wellesley, Mass.: QED, 1990.

Project Tracking and Review

What you measure you know.

Step 3: Project Tracking

Project tracking has one major objective: to determine whether the project is "in control,"—that is, meeting agreed deadlines, quality, costs and so on—or "out of control." As soon as the project has slipped out of control, project replanning, which can include renegotiation of the Business Case and technical specifications, should occur immediately. This tracking process is most simply achieved by a combination of formal tracking procedures and regular team meetings.

The initial focus of project tracking is to review the status of the Business Case to determine any actual or potential variations. Any variation of the Business Case and, in particular, of the scope, objectives, and quality should be subjected to the formalized change control described later in this chapter and in Chapter 6.

Given no major changes to the Business Case, a secondary focus for project tracking (provided that the project plan itself is realistic and accurate) is the comparison of the number of tasks completed with the number of tasks planned for completion, at any given point in time.

Project tracking is therefore dependent on task tracking. Whereas task tracking is undertaken by each team member working on the project, project tracking is made by the project manager based on the actual task effort supplied to him or her by team members and stakeholders using the project plan as the benchmark. Most scheduling tools provide the capability of entering actual effort against the estimated effort.

Provided that the "10–20-day" rule (detailed in Chapter 3) is followed, for purposes of both project and task, tasks must be regarded as either complete *or* not complete; "almost complete" tasks counted as complete will give an inaccurate picture. This approach simplifies project tracking and avoids the 90 percent complete syndrome wherein a task remains almost complete for a period of time. This is called the *Zero-Hundred Percent* technique.

There are other methods of tracking the completion of tasks. One technique commonly used is the *Linear Progress* approach where the percentage complete is calculated from the actual duration versus the estimated duration. If a task was estimated at 20 days duration and 10 actual days have been spent, then the task is 50 percent complete. A variation of the Linear Progress technique is a subjective evaluation of the worth of the actual effort. For example, although 10 days of 20 have been spent, the person undertaking the task subjectively assesses that it is 70 percent complete.

A more complex variation of this technique is *Earned Value*. This technique, involving a series of calculations based on actual versus estimated or budgeted work, is described in more detail in Lois Zell's book [op cit]. Most project scheduling software will calculate Earned Value based on the entering of actual effort through the tracking process. While these and other techniques are common, Thomsett Associates prefers the Zero-Hundred Percent model.

While project tracking is typically undertaken on a weekly or biweekly time frame, it should be emphasized that, as soon as team members or stakeholders realize that they will not meet their task deadline—that is, they are out of control—they should notify the project manager so that the requisite corrective action can be taken. Clearly, this is vital for all tasks on the *critical path* of the project. For noncritical path tasks, this action is required only if the change exceeds the available float for the task.

An essential model for task tracking is a Gantt chart for each person on the project (see Figure 4.1). Most PC-based scheduling tools will provide this chart. These charts provide each team member with a clear picture of their individual work effort, while the overall project Gantt chart provides each team member with a common vision of how the effort of all team members combines in the project. They also are the most useful basis for tracking.

The final focus of project tracking is to collect data to assist in costing and in the creation of an estimating history. This involves project team members recording actual *and* elapsed time spent on the various phases/tasks of the project.

The actual work effort and actual elapsed duration spent on a project phase/task should be recorded daily by each project team member on a project/task tracking document.

This information is required to assist in collecting an estimating history for future projects and, in some cases, for the billing of costs to the project's clients. It is emphasized that it is not a time-keeping or personal evaluation document. It would be quite legitimate for no work to be done on an activity during a day.

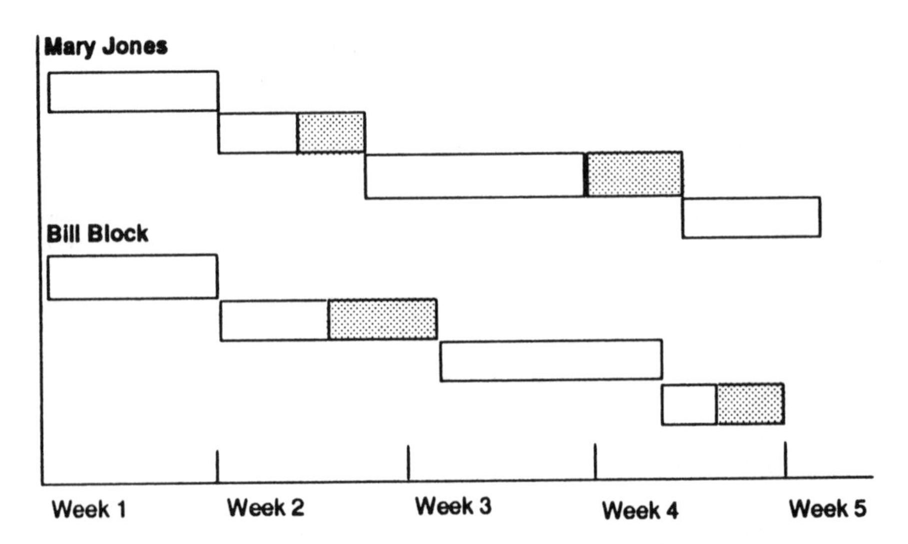

Fig. 4.1
Individual Gantt Chart

It is not necessary to "balance" the number of hours each day or to ensure that the entire eight hours of the day are accounted for.

Remember that project tracking tracks and monitors *projects*, not people.

Using this approach the project manager can then assess the accuracy of the estimated work effort versus actual work effort, and estimated elapsed duration versus actual elasped duration. Project and task tracking should occur on at least a fortnightly basis throughout the project and for high-risk or small projects should occur at least weekly.

The project manager and team members may also wish to track other items such as production time, or activities such as meetings, defect repair, computer usage, travel costs, and so on. Users or clients should be encouraged to track the various activities assigned to them. It may be appropriate for this tracking to be reported to the project manager also because of the impact on the total project cost and effort.

Step 4: Project Reporting

The format and timing of project reporting will depend on each project and organization standards. However, the essential information that should be forward to Steering Committees, Information Systems executives, and senior user area management is:

- The state of the project—is it still proceeding to plans or not?
- If not, what is the revised situation and causes for the variation?
- What actions have been taken by the team to solve any problems?
- What alternative scenarios are available?
- What actions can be taken by senior management?
- Revised or updated Project File (see Chapter 5).

This reporting is also covered in the chapter on Steering Committees (Chapter 6.)

In addition, project reporting would typically involve an aggregation of actual costs to date for the project and, in organizations using cost/recovery and chargeback procedures, the production of invoices for clients and standard charges to the various project service provides such as operations and network.

Control of Project Change and Variation

As discussed in Chapter 1, despite the best of intentions it is inevitable that the need for change will occur some time before implementation. Without a rigorous and negotiated process to evaluate the need for and impact of changes, the Business Case and associated project plan will most likely be disrupted to the extent that drastic rescheduling of tasks or of the entire project will be required. In this context changes can be internal or external:

- *Internal* changes are those that arise during project development due to misinterpretation of requirements, errors of principle or fact, estimation errors, project team member changes, invalid logic, or technical issues that could not be foreseen during the initial planning of the project.
- *External* changes arise through user or client department policy decisions, oversights, new ideas, requirements of other projects, and the like, which cannot be construed as being part of the original system specification.

Although it is likely that an internal change will almost always be accepted as being essential, for control purposes, both internal and external changes must be treated in the same manner.

Control of changes involves three steps: request for change, evaluation, and decision:

- *Request for change.* All requests for change must be in writing no matter what the source; otherwise control will be lost. The requirement is for a brief memorandum addressed to the project manager, which must include the originator's name, date of request, description of the problem addressed, description of the change, and justification for it. It may be necessary to raise a new User Service Request or whatever formal procedure exists for requesting new projects for such changes.
- *Evaluation.* The project manager, liaizing as necessary with other people such as project team members or user professionals, will evaluate the change. This is normally achieved through the convening of a RAP session as described in Chapter 3.

Evaluation should cover such points as the following:

- Is the change really justified?
- If justified, is it essential that it be made at this time or could it or another feature be deferred until the Post-Implementation Review phase at the end of the project?
- Does the change alter the Business Case of the project?

- Which tasks—whether completed, in progress, or to be commenced—are affected?
- What is the estimate of duration and work effort required to implement the change?
- Will it require rescheduling of the project and/or extend the completion date of the project?
- Will it require additional resources to carry out?
- Does the change impact across subprojects or systems?
- Does the change require an alternation of the project development strategy?
- Does it alter the complexity and risk of the project?
- What risks are involved, whether the change is implemented or not implemented?

The results of the evaluation and impact analysis should then be added to the memorandum requesting the change.

- *Decision.* Assuming the project manager has no doubt that change should be made at this time, and provided it will not require additional resources, alter the complexity, change the Business Case, and/or extend the completion date of the project, it can be accepted. However, if one or more of these conditions is not met, reference must be made to the Steering Committee for further consideration.

If there is some doubt, or if the change is extensive, the project manager should call a meeting of interested parties including the requester and evaluator of the change. This meeting should discuss all aspects involved and come up with a recommendation to proceed or otherwise. This recommendation, in turn, may need to be referred to system and/or user department management for confirmation and to the Steering Committee for approval.

Decisions not to proceed must be communicated in writing to the requester. Furthermore, the proposed change may be scheduled as a planned enhancement following successful system or product installation. Such decisions must be considered as final for the duration of the project to avoid wasteful reevaluation of old issues. Decisions to proceed may also be communicated in writing if considered appropriate.

Following a decision to adopt the change, the project manager takes the necessary steps to put it in train. This will include a new RAP session (comment: real-time planning revisited).

It must be carefully noted that, whenever the project moves "out of control" as a result of either external or internal change, the project manager must invoke a cycle of project planning. In cases of severe change, the Feasibility Study phase (refer to Chapter 2) will need to be repeated.

For most projects, there is no such thing as a small change.

A Note on Assumptions

During the RAP session detailed in Chapter 3, a number of assumptions are made regarding the project. For example, while planning the project, the project manager may assume that suitable accommodation is available and that clerical support people are to be allocated to the team.

These assumptions are partly documented in the Risk Assessment model (see Chapter 8). Some are documented in the Quality Agreement (see Appendix B), and some are documented in the stakeholder agreements (see Chapter 5). Any change in project assumptions must be treated to the same change control mechanisms as changes to requirements, schedules, and other key Business Case components.

5 Project Agreements

If it's not written down, it doesn't exist.
Gerry Weinberg [1981]

At the end of the initial project planning sessions, which generally precede the Feasibility Study or Terms of Reference phase and the Systems Requirements phase, the project manager and team will have developed a series of documents that describe the managerial and business aspects of the project.

Project File

These documents include the Business Case as described in Chapters 1 and 2:

- Project overview.
- Project scope.
- Project objectives.
- Project returns.
- Project costs.
- Project development strategy.
- Project risk assessment.
- Project development schedule.
- Relevant legislation.
- Stakeholders/key groups.
- Project staffing.
- Interrelated projects.
- Assumptions and constraints.
- Project/product quality.

In addition, the team will have developed task lists, estimates, assumptions, detailed schedules, administrative procedures and standards, and other working papers. These papers comprise the managerial issues associated with the project and should be assembled into a single file called the Project File or Project Management File. It is important to keep these documents separate from the technical documents associated with the project such as functional requirements.

Stakeholders/Key Groups

Traditional project management approaches were focused on a single project team, a single project manager and a single user group. Contemporary projects are more complex, and it is typical for a project to involve many disparate support, consultancy, and user groups. These external groups are also termed stakeholders or key groups.

For example, a computer project in a large organization may involve the following groups:

- Other project managers of related projects.
- User area Business Manager/s.
- Project teams (one for each subproject).
- Steering Committee.
- Senior management group.
- A number of unions and staff associations.
- Various user/client groups.
- Federal and state government departments.
- Multiple vendors and software suppliers.
- Internal technical support groups such as Data Base Administration, Communications Network, Operations, etc.
- Internal Audit and Quality Assurance.
- Related projects.

In other words, in developing a new system or enhancement the project manager is required to consult and act as a "contract manager" with numerous external groups, projects, or subprojects and a number of support, development, and technical "subcontractor" areas.

The role of these external groups in contemporary projects can vary from a passive review role to a fully involved active project role. To reflect this possibility, the project manager should structure the external groups involved in his or her project into at least six levels of impact or involvement. These are:

- *Level 1:* Groups/organizations outside the project manager's organization who need to passively review or audit the project and its deliverables. Level 1 groups include Enternal Audit, Government or industry standards areas and indirect users.
- *Level 2:* Groups/organizations outside the project manager's immediate area who have a minor participative or consulting role in the project. Level 2 groups include Internal Audit, quality assurance areas, and project development support groups such as methodology and methods.
- *Level 3:* Representatives of external systems or projects that either send or receive data into or from the project. Level 3 groups include computing representatives of related systems and projects and users form those systems.
- *Level 4:* Groups and projects who are within the organization but outside the project manager's immediate organizational area and who have a major par-

ticipative or development impact on the project. Level 4 groups include internal technical support groups such as Data Base Administration, Operations and Network experts.

- *Level 5:* Groups and projects outside the project manager's organization that have a major participative or development impact on the project. Level 5 groups would include vendors, contracting staff, external consulting groups, Trade Unions, Staff Associations, and any people in other organizations involved in interrelated or dependent projects.
- *Level 6;* Client groups who are the primary sponsors/clients of the project. Level 6 consists of management and key people from the project's direct client areas and the Steering Committee: The project team are always treated as Level 6 stakeholders.
- It is common for each of the external groups to have different perceptions of the project's objectives, scope, risk, quality, and other project management concerns. Therefore, it is essential that a least Level 3, 4, 5, and 6 groups are actively involved in the project planning, tracking, and review process. Ideally, representatives from each Level 3, 4, 5, and 6 group should be part of the team-based RAP sessions described in Chapters 2 and 3. It is essential that representatives of Level 5 and 6 groups *at least* are present at the planning sessions and that Levels 3 and 4 stakeholders are involved via an active review process, such as Freedman and Weinberg's [1982] Technical Reviews and Walkthroughs.
- Level 1, 2, and 3 group representatives should be involved if possible, but, at a minimum, they should be involved in review and approval of the project plans and Business Case.
- In the case of related projects, it is useful to distinguish between the type and nature of dependency. In terms of dependency type, the project being planned may be dependent on another project, be interdependent with another project, or have other projects dependent on it. The nature of dependency can include:

- *Data.* The project shares data with another project/s.
- *Function.* The project shares common functionality with another project/s.
- *Objects.* As discussed in Chapter 11, the projects may share data and function as units or objects.
- *Staff.* The project shares staff with another project/s.
- *Technology.* The project shares technology, that is, one project installs the technology that another requires.
- *Funding.* The project shares funding arrangements, that is, one project funds another.

Depending on the nature and type, the project manager may require specific people involved in assisting in managing the other projects as stakeholders. For example, a data dependency is handled by a trained data administrator and a funding dependency by a financial expert/account.

Project Agreements or Charters

As a result of the criticality of external groups, project managers will find that their ability to meet project deadlines, quality requirements, and deliverables specification are dependent on people from Level 3, 4, 5, and 6 external groups.

Prior to the commencement of the project, when the project is dependent on external group people, the project manager should obtain a *formal* commitment from these external groups that they are prepared to release their resources in accordance with the proposed project schedule. In many cases, this requires a formal approval of the Project File, the Business Case, and the proposed schedules by key representatives of each Level 3, 4, 5, and 6 external group. In the case of Level 5 groups, such as vendors or commercial providers of resources, this would also involve formal contracts of service. These formal agreements are documented as the Project Agreements or Project Charter.

Typically, these agreements document the following:

- The type of support/service required.
- The timing and costs of the services.
- Specific performance measures, if applicable.
- Fallback or contingency arrangements.

Any conflict of priority or requirements among the various external groups should be raised with the Project Steering Committee for resolution (see Chapter 6).

In summary, for many projects a series of formal or semiformal agreements must be drawn up between the project manager and the various external groups involved in the project. These Project Agreements form the basis of the change control process described in Chapter 4. Any formal agreements should be added to the Project File.

As discussed in Chapter 9 and Appendix B, there should also be a formal Quality Agreement negotiated for the project.

Staffing Agreements

A common mistake made by project managers is the assumption that people are equal in skills and experience. Research by Capers Jones [1986], DeMarco and

Lister [1987] and others has revealed what most experienced computer people have intuitively understood for many years: There are differences between individual productivity of at least 2:1 and, in some areas such as analysis and design, 10:1 or greater.

Given that the process of project planning often requires the project manager to assume that specific skills will be available, it is simply pragmatic for the project manager to carefully specify the required and assumed skills base required for the project. This is especially the case where the project manager is not aware of the specific people who'll be allocated to the project.

Some organizations have developed and maintained a skills model for their computing people and this should be used for documenting staffing assumptions.

A typical skills definition would include the following details:

- The specific skill such as COBOL or data analysis.
- The skill level required, such as Levels 1–4 (where Level 1 means understanding the skill and able to practice under supervision, and Level 4 means an expert practitioner capable of innovative research).

Once these assumptions are documented in the Staffing Agreements, should people be allocated with different skills than those in the Staffing Agreement, then the project manager has a solid case for renegotiating estimates and project risk. Figure 5.1 provides a sample form that can be used for documenting staffing and services.

A Final Note on Agreements

It is important to understand that the various project agreements negotiated between the project manager and the stakeholders should be treated as "two-way" contracts whenever possible.

For example, when the project manager presents the Business Case and associated project plans to the Steering Committee, the Business Case is in effect a contract between the Steering Committee, the project sponsor, and the project manager. It is a contract that commits the project manager to ensuring that the Business Case is maintained and managed to achieve the agreed-upon scope, objectives, and quality within the agreed-upon deadlines and costs.

However, certain issues—such as unresolved high-risk factors (see Chapter 3 and 8), boundary disputes, and service issues with stakeholders—require a return commitment from the Steering Committee to assist the project manager in managing these risks and disputes. This concept is explored further in the next chapter.

In other words, while the project manger is responsible for the management of the development process, the complexity of the stakeholder environment in most projects means that the project manager must be able to depend on the

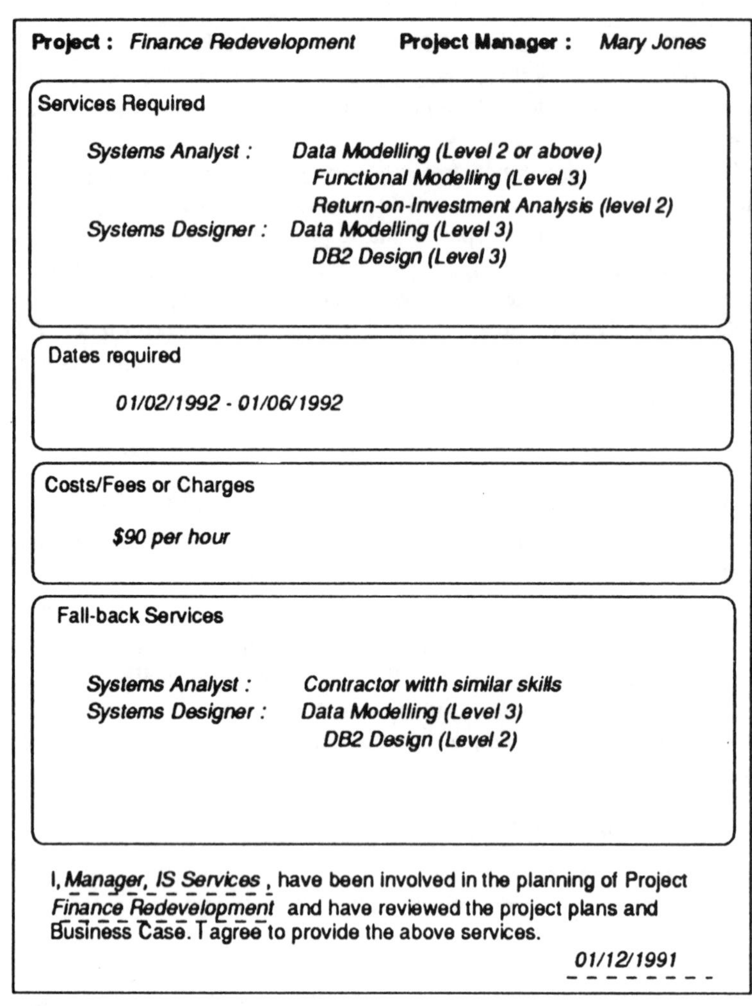

Project : *Finance Redevelopment* **Project Manager :** *Mary Jones*

Services Required

Systems Analyst : *Data Modelling (Level 2 or above)*
Functional Modelling (Level 3)
Return-on-Investment Analysis (level 2)
Systems Designer : *Data Modelling (Level 3)*
DB2 Design (Level 3)

Dates required

01/02/1992 - 01/06/1992

Costs/Fees or Charges

$90 per hour

Fall-back Services

Systems Analyst : *Contractor witth similar skills*
Systems Designer : *Data Modelling (Level 3)*
DB2 Design (Level 2)

I, *Manager, IS Services* , have been involved in the planning of Project *Finance Redevelopment* and have reviewed the project plans and Business Case. I agree to provide the above services.

01/12/1991

Fig. 5.1
Staffing Agreement Form

project sponsor and Steering Committee members to assist in ensuring that the various project agreements are honored.

To summarize, the Business Case and the other project agreements are based on the project managers being assisted by the senior management of the organization in certain areas such as managing risk and stakeholders.

It is inevitable that in contemporary projects, conflicting demands and priorities will converge on the project manager and the stakeholders. The Steering Committee must act as a forum for dispute resolution when these conflicts arise.

References

T. DeMarco & T. Lister, *Peopleware*. New York: Dorset House, 1987.

D. P. Freedman & G. M. Weinberg, *Handbook of Walkthroughs, Inspections and Technical Reviews,* 3rd ed. Boston: Little, Brown & Co., 1982.

C. Jones, *Programming Productivity*. New York: McGraw-Hill, 1986.

G. M. Weinberg, ''Technical Review Workshop,'' Canberra, 1981.

6 Steering Committees

Steering Committees should solve problems—not create them.

As discussed in Chapter 1, the involvement of senior management in computer-based projects is essential in ensuring that the approval, development, and support of information system projects reflect the organization's strategic direction and that the allocation of people, capital, and equipment involved in projects is done within a broad organizational perspective.

Also as outlined in Chapter 1, in all projects there are two control systems.

- The *managerial* control system, which covers objectives, people, time, costs, stakeholders, and other business issues.
- The *technical* control system, which deals with the specific issues and concerns associated with the equipment/technology and the deliverables of the project.

It is generally accepted that the technical control system is delegated through the project manager to the technical experts and that the senior management focus on objectives, people, time, risk, and priorities, that is, the managerial control system.

Like project managers, Steering Committees should become involved not in resolving specific technical issues, but rather in resolving the *impact* of those technical issues on the project's objectives, costs, and so on. The focus of Steering Committees must be on the Business Case and its associated priorities, strategies, deliverables, and plans. To affect these factors, senior management must be informed on the status of the project components of objectives, people, time, risk, and quality via the reporting process detailed in Chapter 4.

The content of the project reports are critical in ensuring that the Steering Committee can focus on the correct issues.

Given that most Steering Committees are dependent on the project manager and project reports for their information, then by ensuring that the project reports concentrate on the management issues of the project, the project manager can create a Steering Committee environment that "adds value" to the project management process.

Steering Committee Functions/Responsibilities

There are four major functions of Steering Committees:

- Selection and approval of projects.
- Monitoring and review of projects.
- Assistance to projects when required.
- Resolution of project conflicts.

Approval of Projects

As detailed in Chapter 2, the approval of projects is a two-stage process. The first stage is the identification of potential projects (via new legislation, Strategic Planning, or normal business initiatives) for further investigation.

Following an initial planning session, these projects would be approved for the development of a full Business Case via a Feasibility Study (see Figure 6.1) For minor projects, a Steering Committee would probably not be formed, and the project sponsor would undertake the role of the Steering Committee.

In larger organizations, there is generally a permanent executive-level committee, which overviews and manages all information systems projects from a strategic perspective and which delegates specific control of individual projects to project-level Steering Committees.

At the end of the Feasibility Study Phase, the Steering Committee receives a Business Case report as described in Chapter 2.

On the basis of the Business Case, senior management determines whether to proceed with further development. Should the project be suitable for further development, the Steering Committee should prioritize the objectives, approve the budget, specify mandatory outputs/deliverables, schedule and budget constraints, and determine what actions they are prepared to undertake to minimize the risk of the project, that is, to negotiate a two-way contract with the project manager.

Senior management should also determine the degree of delegation to the project manager and the major review points during the project cycle.

Monitoring and Review of Projects

Depending on the size, risk, and organizational impact of the project, the Steering Committee will monitor and review the project at various stages using the project reporting process introduced in Chapter 4.

At a minimum, for projects proceeding according to the approved Business Case, the Steering Committee should review the project at the end of each phase of the project development cycle. However, the monitoring and review process may be required more frequently for projects that are subject to a high risk of change, or projects with a high degree of organizational impact, for example. Further, if the project is using more complex development strategies such as Hybrid, Rapid Application Development, or Fast Track, the Steering Committee review process will be more dynamic and flexible than the traditional "phase-end" review as the traditional distinction between phases is lost in these strategies. Typically, the Steering Committee review process in the more complex strategies is determined by deliverables such as prototype completed, JAD sessions concluded, and so on.

The process of monitoring and review should focus on the Business Case and any variations in its key components, such as objectives, risk, cost, returns, and quality.

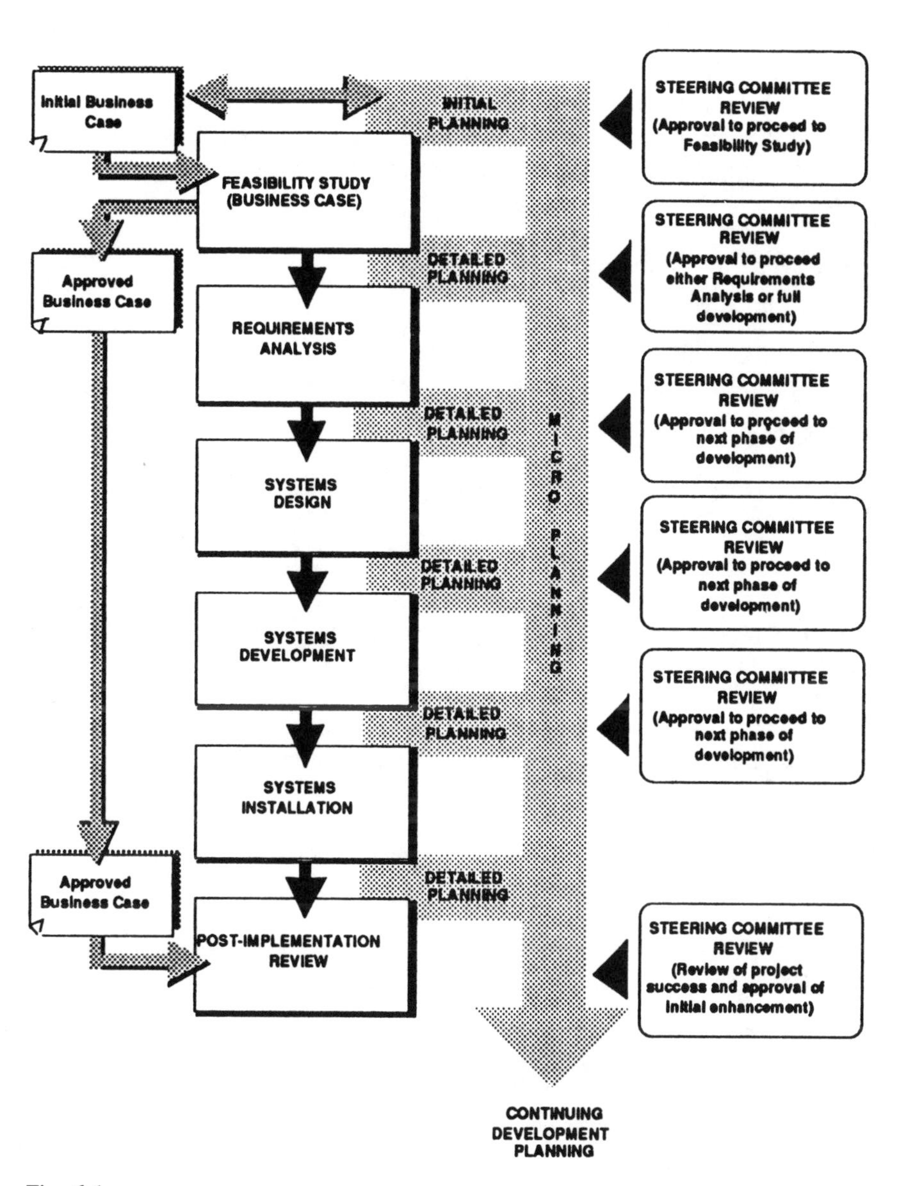

Fig. 6.1
Steering Committee Reviews

The Steering Committee is the only group to approve changes to the Business Case. As a rule, the management control of the components above should *not* be delegated to the project manager. Should any variation in the original Business Case be proposed for the Steering Committee's consideration, the following information should also be provided for the Steering Committee meeting:

- The nature of and reason for the variation.
- Effect of the change.
- Revised Business Case (including alternative project plans).
- Actions taken by the project manager to contain the change.
- Suggested actions for the Steering Committee to consider.

It is typical that, during the review meetings, accumulated costs and other related issues are also reviewed.

Assistance to Projects

In the case of changes to projects, the Steering Committee's function is to assist the project manager in resolving the change and proceeding with the project.

In many cases, changes to projects are beyond the project manager's control, and assistance from the Steering Committee in resolving the change is a valid function for senior management.

While it is required for the Steering Committee to understand the reasons for the variations in the project, it is more important for them to focus on resolving the situation and ensuring that the project continues to be developed within their control.

Resolution of Project Conflicts

A consistent theme throughout this handbook is that, in contemporary projects, the project manager is acting as a "contract manager" who will require a number of "subcontractors," that is, experts from areas outside the project manager's direct managerial control. For example, the project manager may require data base, personnel, vendor and Regional Office people to assist in various phases. In addition, as described in Chapter 5, the project would typically have a number of external stakeholders as well.

In this situation, it would be common for a project manager to become involved in conflicts regarding the allocation, quality, and level of commitment for people involved in the project on a "subcontractor" basis. These subcontract arrangements should have been documented in the Project Agreement described in Chapter 5.

In this situation, the Steering Committee can provide valuable assistance to the project by resolving these inter- and intraorganization conflicts.

Other potential conflicts that would require assistance from Steering Committees are conflicts in the priority of objectives or deliverables, external attempts to alter the Business Case by a particular user professional area, boundary or scope disputes between other project managers, and stakeholders issues relating to Staff Associations, unions, and so on.

Steering Committee Meeting Structure

While the structure and agenda of Steering Committee meetings vary from project to project, all Steering Committee meetings should directly address the following topics or agenda:

The status of the project—has the Business Case altered since the last meeting?

If so, what changes are there to:

The scope of the project?

Objectives?

The priority of the objectives?

The project development strategy?

Risks?

Benefits or returns?

Costs?

Project plans/schedules?

Quality agreement?

Staffing agreements?

- What actions has the project manager/team taken to resolve the issues?
- What actions can the Steering Committee consider to assist in resolving the issues?
- Who is responsible for follow-up?
- What caused the changes?
- Any future/expected issues?
- Other items as required?

Should a specific technical issue require detailed discussion and resolution, it is recommended that a specific Technical Review meeting is scheduled following the normal managerial Steering Committee to avoid senior management becoming distracted from the managerial control of the project.

A Final Note on Steering Committees

Given that the members of Steering Committees are usually senior management and that senior management are usually busy, it is typical that project managers are under considerable pressure to simplify and summarize the issues in a project—especially a complex or large project.

Despite this understandable pressure to reduce critical issues to "one page or less," the project manager must ensure that the members of the Steering Committee are given sufficient information to ensure that they are fully aware of the status and issues of the project. Otherwise, Steering Committees are reduced to a "rubber stamp," which unfortunately is the fate of many Steering Committees.

The use of common formats, desk-top publishing technology and graphics, together with patient and plain language explanations of the issues, can often help the members of Steering Committees come to a more realistic understanding of the necessary complex issues of most projects.

Unfortunately, complex projects require complex models.

7 Return on Investment Approach

People don't pay for technology.
They pay for what they get from technology.
Peter Drucker [1985]

Ranking and Presenting Objectives

Objectives are the *critical* definition of a project. It is therefore paradoxical that they have existed in a twilight zone between systems analysis and project management, with neither function taking responsibility for them.

The derivation and refining of project objectives is a task that should be shared between the project manager and the system or business analysts and the project's clients.

It is typical for high-level objectives to be stated before a project commences. One of the key deliverables from the Feasibility Stage Phase is a refined and more detailed statement of objectives for the project. These objectives drive both the project management process and technical project development cycles.

Generally, a business or system analyst conducts, in the course of investigating the current situation, a series of interviews with the client's people who are involved in specifying the system's requirements (Level 6 external group) and with representatives of all other external groups (Levels 1–5). These interviews are in essence an ''audit'' of the objectives for the project, which typically includes the correction of any existing problems and new innovations reflecting strategic or business-level plans.

At the end of this objective's ''audit,'' the team should have a good candidate list of objectives, comprising the problems and opportunities that the project must address.

Now a different set of problems will be apparent. Firstly, the objectives that have been identified will usually be at different levels in the classic organizational hierarchy. For example, there may have strategic-level objectives gained from interviews and reviews with senior management sitting right next to very detailed system-level objectives gained from the clerks' performing a particular operational process.

In a typical project, at least five levels of objectives can be defined:

- *Strategic*. Objectives that impact on the whole organization and reflect the organization's mission.
- *Management*. Tactical-level objectives that reflect a divisional or group perspectives.
- *Project*. Objectives specific to the project as a whole including noninformation system objectives.
- *System*. Specific objectives in terms of data and function relating to the information system component of the project.
- *Design*. Technology-dependent objectives specific to the physical design of the information system.

The project manager and team must rearrange the objectives into this hierarchy. Experience has been that it is fairly easy to isolate strategic objectives as they impact most areas in the organization. System-level objectives are also fairly easy to isolate because they focus on specifics of the system. Generally, the team knows most of the system objectives from their initial interviews with clients. Surprisingly, the management-level objectives are often the hardest to isolate and evaluate. Design objectives would normally be determined during the development process.

Second, some of the objectives may be physical (design) objectives rather than logical objectives. Physical objectives would typically refer to implementation via computerization or a particular package or manual solution. In other words physical objectives state "how" the problem will be solved in terms of technology and resources, whereas logical objectives state "what" is required and should be free of any implementation or technological constraints. For example, consider an objective such as to put customer records onto DBase IV. A team can always clarify physically stated objectives by subjecting them to the "Why" test. Why is there a requirement for DBase IV or any data base for customer records? The answer is to get faster and alternative types of access to customer name and goods, a logically stated equivalent.

Third, some of the objectives will not be real objectives. False objectives are often statements of end or *constraint*—for example, to implement a cheap system or to make a lot of money or to implement the system by July 1, 1992. This type of objective will fail the Returns Analysis test detailed later in this chapter. They should be ignored until you are considering the costs of the alternative solutions, or documented as Project Constraints and Assumptions within the Business Case.

Fourth, many of the system objectives will have an impact on or have related objectives in other areas of the organization.

For example, it will be no use for our system to provide expanded information to management if they do not have the requisite training and support systems to handle it. In theory, that is a problem for management as an external process to the project. However, in practice, the team should be aware of any related objectives and ensure that, for any changes that they make, the impact on people involved in external areas or projects is suitable to them. Of course, the higher the level of objective in the hierarchy, the wider the related objectives will reach. Again, the role of external groups is essential here in helping to specify and prioritize any objectives relating to their concerns.

Finally, objectives are often politically driven, causing problems especially if multiple areas must agree on the objectives for the system.

A very powerful tool for revealing potential clashes between counterproductive objectives which could lead the project to becoming a victim of political games is *ranking*. For each objective in the hierarchy, the team must ask each person involved in the system and the external groups or stakeholders to rank the system as mandatory/must (essential) or necessary/should (not critical in the short term).

Ideally, all participants should agree on the ranking outcomes for each objective. If not, then it is strongly suggested that, unless the project manager is masochistic or a very good negotiator, he or she raise the area of disagreement with the project sponsor and the Steering Committee for resolution.

Putting this all together, there should be a list of real objectives, each logically stated, ranked as mandatory or otherwise, and placed in a hierarchical structure. The members of Thomsett & Associates have found the structure in Figure 7.1 useful as it also shows where other areas of the organization have related objectives as well (other stakeholders). This is because many management-level objectives are enabling-type objectives; that is the management role is to enable the system to achieve its objectives.

STATEMENT OF OBJECTIVES Ardvark Imports Pty Ltd

AREA / OBJECTIVES	ACCOUNTS	ADMINIS-TRATION	MA
Strategic			
To expand scope of operations into manufacturing	M	M	
To consolidate existing client base	M	M	
Management Accounts			
To improve cash control	M	N	
To provide tighter Account system	M	N	
System Accounts System			
To balance Accounts daily	M	M	
To follow-up O/S Invoices daily	M		
To gather more customer details			

M - Mandatory N - Necessary

Fig. 7.1
A Simple, Integrated Objectives Statement

An Approach to Returns Analysis

Armed with an integrated and reviewed objectives statement, the team can now subject it to a Returns Analysis.

The subject of Cost/Benefit or Return on Investment (ROI) analysis has always been problematical in computing. Part of the problem is that they do not actually belong together and are calculated at different stages of the development process, as shown in Figure 7.2. Another part of the problem, as already men-

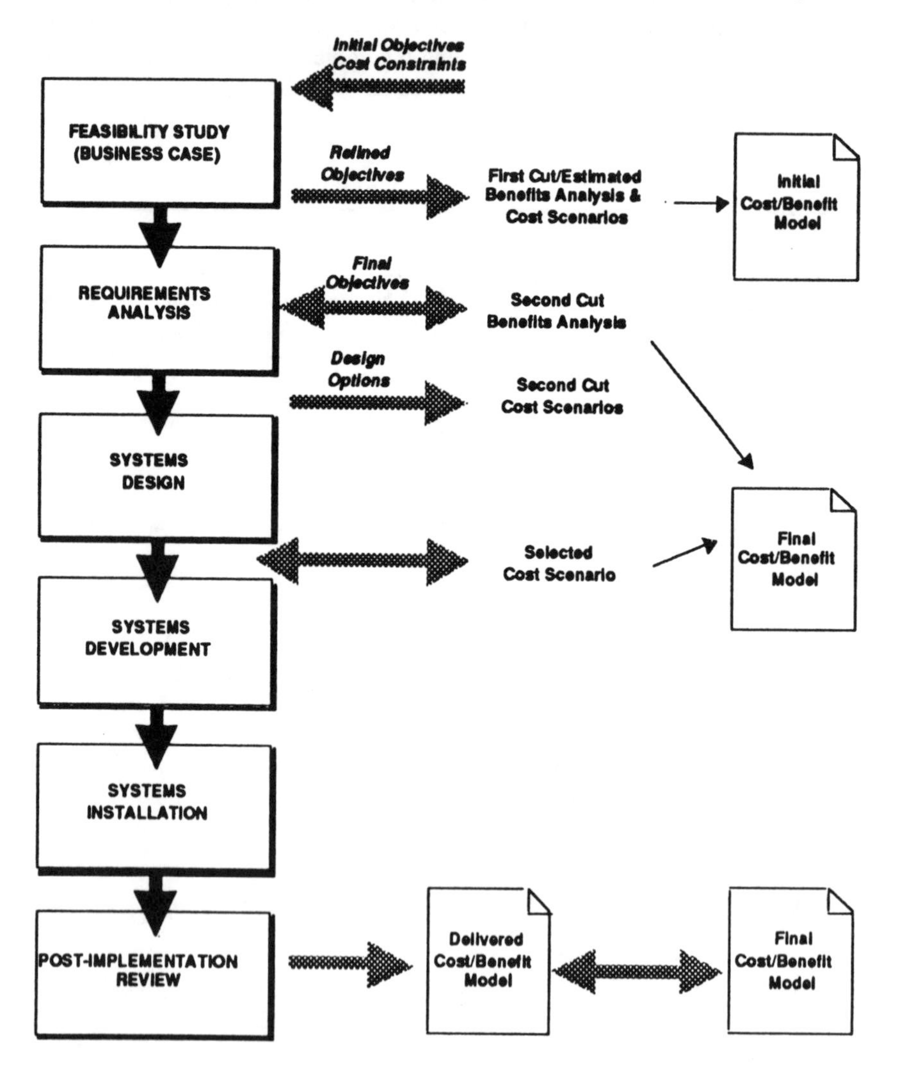

Fig. 7.2
The Evolution of an ROI Analysis

tioned (also indicated in Figure 7.2), is that the estimates of costs and benefits are required at the project justification stage where, in most cases, they are only "ballpark" estimates. Finally, most project managers and systems analysts have only basic education in developing a Cost/Benefit model, and this education is based on traditional capital equipment acquisition or accounting approaches to CBA.

A rigorous approach to both benefits and cost analysis is essential in contemporary project management. Not only does it ensure that the project is justified from both the corporate and financial aspects but, the process of Cost/Benefit analysis detailed in this chapter provides the project manager and team with a clearer understanding of what the project is really about. In other words, Cost/Benefit analysis is a case where *the means is the ends.*

As outlined later in this chapter, benefits or returns can be calculated from the logical objective statement. The costs or investment cannot be calculated until the physical alternatives determined as effective for the new system model have been derived and one is selected. Typically, there will be a number of effective solutions and each will have a different cost.

For a given set of objectives, the returns will be fixed but there will be alternative solutions: fixed returns and variable costs.

A second and very real part of the problem is that it is usually easier to calculate the costs/investment than it is to calculate the benefits/returns. However, the following concepts and approaches do help. Even if the results are not totally satisfactory, the project team will certainly know a lot more about what they are trying to do and they will have further refined the objectives list.

The area of returns/benefit analysis has often been partitioned into tangible and intangible benefits. While recognizing the persistence of this approach, another approach is that the team *ignores* all intangible benefits. Since they are intangible, they cannot be measured, so why take them into account at all? Using the approaches following, most intangible benefits can be converted into measurable benefits.

Objectives will lead to *direct* and *indirect* returns. For example, by delivering a particular service to customers, the project/system will not only improve service (direct or primary), but may attract new customers (indirect or secondary) and this may lead to more revenue (indirect). The fascinating thing about this approach is shown by the story told by Gerry Weinberg in one of his seminars about the child who was taught how to spell banana, but not taught when to stop. The team may not know how far to go before they stop the process of deriving related secondary benefits. For example, having gotten more revenue, the team's company employs more people (indirect), who stop worrying about day-to-day survival (indirect) and who can now support the Nuclear Disarmament Party (indirect), which leads to their being elected (indirect) and the country leading the world to a nuclear-free future (huge indirect return). In action here is the cause-effect dilemma much discussed and beloved by psychologists, anthropologists, and economists.

In summary, only direct or primary measurable returns should be evaluated and measured. Any indirect or secondary returns may be documented to be considered as additional factors, which may influence the project's approval. Further, only one level of indirect returns or secondary returns should be calculated. One useful output from this approach is that the project should only be held responsible for the achievement of the primary or direct returns. Secondary returns will require external groups and stakeholders to take action for their achievement. For example, an office automation project may be successful in implementing the technology, but because management is not supported through education, the secondary benefit of increased management effectiveness is not achieved.

The most useful technique for deriving primary measurable returns is the modified IRACIS method, derived from Gane and Sarson [1978], which involves classifying the primary benefits derived from each objective into one or more of three categories:

- IR—Increase Revenue
- AC—Avoid Cost
- IS—Improve Service

In this approach, each system-level objective is evaluated in turn to see which one of the preceding return categories the objective is attempting to achieve. In many cases, an objective may fall into two or more return categories. In this case, the team can often refine the objective into two or more "finer" objectives. Figure 7.3 shows how this technique is applied.

It is important to note that usually only the Increase Revenue and Avoid Cost categories will be financially measurable. For example, reduction of the number of people required to run a system is an Avoided Cost return. The financial saving can be derived from calculating salaries plus overheads for each saved position. However, the Improve Service category is measurable. For example, it is possible to measure whether the system has improved service by simple interviews or by prior and post measurement. What is essential is that the Improved Service objectives have specific measures defined for evaluating the success of their achievement. For example, an objective such as "To improve management reports" can be restated as "To provide the Manager, Personnel with daily attendance figures by 5.00 P.M." Tom Gilb [1988] provides a very sophisticated method for quanitifying objectives. However, for most projects, it

SYSTEM Accounts System

To balance Accounts daily	*Avoid Costs (management lost time 2hrs/week @ $ 60.00 per hour), Improve Service (Admin. managment daily reports, Advertising information)*
To follow-up O/S Invoices daily	*Avoid Costs (Accounts people 4hrs/week @ $40.00 per hour), Increase Revenue (tighter cash-flow & investment of funds @ 18% p.a. - $10,000 /yr)*
To gather more complete customer details	*Improve Service (Advertising - improved client profile for selecting clients), Increase Revenue (exp. 10% growth from improved client info - $1,000,000 /yr), Match Competition (better client relations)*
To provide Sales Tax returns in new format (Sales Tax Legislation 1986/3)	*Avoid Costs (penalties of $10,000)*

Fig. 7.3
Returns Analysis

would be sufficient if the objectives are specific and a measure is determined as shown in Figure 7.4.

To improve management information	To provide Accounts with client category updates by friday p.m.
	To provide client name, balance and rates by 5.00 p.m. daily
	To give management on-line query facilides to client data by client category

Fig. 7.4
Refining Loose Objectives

Some other considerations may be useful in applying this technique:

- Improve Service should result in Avoided Costs or Increased Revenue as secondary returns. It is very important to attempt to measure this secondary return of Avoided Cost or Increased Revenue to the recipient of the improved service. For example, by producing reports daily instead of weekly (Improve Service), we may save management one hour's work per day (Avoided Cost). This will help prevent Improve Service category objectives becoming a catch-all for "intangibles."

- If the objectives do not fall into the IRACIS categories, they are not real objectives. They will probably need to be either eliminated or broken into more specific objectives.
- There is a whole category of objectives dealing with quality of working life, job design, and job satisfaction. In some organizations, an objective such as "To improve morale of staff" may be accepted as a valid objective. While these may break down into measurable objectives as shown in Figure 7.4, they do require special attention.

In summary, real objectives must pass the following tests:

- They must fit into one or more of the IRACIS types.
- They must be logical—What, Why, Who, When, Where, not How.
- If we take a real objective away, the project must be different in scope and returns.

It is common that objectives will change throughout the project as new or altered business requirements emerge or as existing objectives are further refined. Also, the various project implementation alternatives may give rise to new objectives. For example, if it is decided to automate the system, it may be possible to add new features to the requirements at no or minimal cost. The use of a data base to store client records could give the ability to merge client records with a form letter layout with public relations information.

So, prior to full costing of returns and investment, it may be necessary to revisit the initial objectives statement, update it, review it with the key groups of the system, and subject the revised objectives to a formal change control process (Chapter 4). As shown in Figure 7.2, this process would normally be undertaken during the Systems Analysis phase.

Of course, the reverse may also apply. As a result of the additional analysis, what appeared to be a reasonable set of objectives may not be achievable (physically). Should this happen then there will be a need for renegotiation and reanalysis of a degraded or reduced objectives statement.

What is important is not to assume or anticipate even small variations and to recalculate returns prior to costing the alternatives. The objectives statement drives the subsequent system and the project management process. Do not change it without reviews with people from the key stakeholder groups.

If the project team loses control of the objectives, they have lost the control of the project as well.

Cost Models

While most sources agree on the basic frameworks for costing, there is considerable debate about the level of precision and detail of cost models.

Take people as an example. Some simple models assume one person's cost equals their annual or average salary. Other more complex models add management overheads, on-costs, and actual days of work. More advanced models factor in accommodation, cleaning costs, air conditioning, and so on. One complex model also includes education as an amortized component affected by staff turnover rates and average length of employment.

The model used in this handbook is one of the medium complexity models. It has been modified from an comprehensive document produced by the Australian Department of Finance [1991].

This model basically starts with salary and then adds a set of factors to derive a total per year cost.

Salary	100%
Pension/below-the-line	29%
• Superannuation	
• Shift, overtime	
• Separation pay	
• Per diem, etc.	
Administrative expenses	48%
• Cars/taxi-hire	
• Telephones	
• Copiers, fax machines	
• Supplies, etc.	
Accommodation	24%
Corporate support	54%
• Executives/managers	
• Clerical support	
• Typists/DPOs, etc.	
Total per-year cost	255% (salary)

To derive a per-hour cost, should that be required, it is important to exclude weekends and public holidays. These figures vary by 200–220 days per year at 6–7 hours per day are reasonable guidelines. The hourly cost of a person on $30,000 per year then works out at between $40 and $50 (excluding use of computers, networks, etc.).

The costing of equipment—computer software, hardware, supplies, forms, and so on—is generally easier as long as it can be assumed that the equipment or capital investment is dedicated to the project and is paid for within one year. Should the project's personal computer, for example, be used for other systems or by other people, it will be required to estimate the percentage of use by each system to derive a project-oriented cost. Of course, it is also possible to write off the shared use by absorbing the cost totally into one project. Many organizations provide a standard costing model for equipment usage such as CPU usage time and network charges/transaction.

The Finance Guidelines also add maintenance, replacement supplies, resale or residual value, and cost of site and site-related components—false flooring, air conditioning, back-up power supplies, security mechanisms, and so on. Should the project be required to add these cost factors, then read the guidelines or discuss them with the organization's operations people.

To summarize them, there are two major cost categories: people and equipment. However, there are additional issues associated with the costing of projects.

Other Factors—Time, Time, and Time

Time plays a significant factor in the project development costing process—in three ways.

It is important to separate costing for ROI analysis from accounting for taxation or financial reports. This has always been confusing to some people but the Taxation Office or Internal Revenue organizations allow capital equipment to be depreciated at set rates. However, for the project it is possible to write off or "depreciate" the capital equipment at a different rate. Welcome to the wonderful world of creative, contemporary accounting.

The first way in which time affects ROI is in the concept of capital and recurrent costs. Consider the acquisition of a computer. The company can buy it outright or lease it. In the purchase option, the company (or project) will incur a one-off capital cost this year, which may or may not be amortized or written off over a varying period of time. The period of amortization varies from company to company and can range from two to ten years. In addition, the rate of write-off can vary over time depending on your accounting procedures, such as 40 percent first year, 20 percent per year for the next three years.

In general, the longer the period of amortization, the lower will be the impact of the cost on the project's ROI. For example, $3,000 over ten years (excluding inflation, etc.) is $300 per year. For a cost-returns analysis this looks better in the short-term than a three-year-write-off of $1,000 per year. In general, the purchase of finished product (such as a software package, consultant-developed software, etc.) is considered to be a capital investment.

Should the company or project lease the computer, the project will have a recurrent cost per year of the lease payments. *Recurrent costs* are those costs that are expected to be incurred over a number of years. This concept also applies to people. The use of in-house people is classified as a recurrent cost, while the cost of a consultant involved in a three month contract could be considered as a capital cost.

The second and more complex way in which time affects a project's costing is in the concept of Present Value. This concept recognizes that $1.00 spent this year is worth more than $1.00 spent next year (assuming inflation is here to stay).

Similarly, a benefit or return of $1.00 this year is worth more than a $1.00 return next year. So to compare the benefits or investment anticipated in three years' time with returns and costs for this year, a discount rate must be applied to the returns and benefits in the second and third year to bring them back to today's values or Present Value. The general equation and an example for this is shown in Figure 7.5.

The value of discount rate again varies with organization and level of detail. A simple and useful discount rate is the inflation rate which, in Australia, is 7–10 percent per annum.

More complex discount rates include consideration of the level of interest rates and opportunity cost. For example, if a person can obtain 15 percent per annum by placing money in a bank, then by placing the money into the project, they should expect at least a 15 percent return from the project, which takes the discount rate to 15 percent.

For example, the Australian Department of Administrative Services provides for a discount rate at three levels of risk:

- *Riskless rate:* The yield on the longest running Treasury bond.
- *Medium risk:* 150 percent the riskless rate.
- *High risk:* 200 percent the riskless rate.

Some companies have discount rates as high as 25%. This more complex discount rate is also called the Marginal Cost of Capital (MCC).

The final way in which time affects our cost calculations is by means of Price or Cost Movement Factors. Both our people and equipment components have a cost movement over time. Hopefully, people will earn more next year and equipment should cost less next year (in actual, not Present Values). So before the team calculates future annual costs, they need to increase or decrease the

Present Value = Annual Value for year $n/(1=i)_n$
where n = year
i = discount rate
$1/(1+i)_n$ = discount factor

Assume $1000 is to be received in benefits in 5 years time and we apply a discount rate of 16% p.a.

Present Value = $\$1000/(1+0.16)_5$ = $1000/2.1368
Present Value = $468

Fig. 7.5
A Sample PV Calculation

actual current year value by means of Price Movement Factors prior to calculating Present Values. For example, people could have a PMF of +10 percent per annum and equipment of −10 percent per annum.

Once Present Values have been derived for both the costs and benefits, there are four common ways of showing return-on-investment.

- *Net present value* is derived simply by taking the sum of the Present Value of the costs from the sum of the Present Value of the Returns. If the result (Net Present Value) is positive, then the project is viable.
- *Return-on-investment* involves dividing the Net Present Value by the sum of the Present Value of Costs and deriving a percentage.
- *Internal rate of return (IRR)* is derived by calculating the discount rate required to make the Returns equal to the Costs (both calculated Present Value terms) over the project's agreed pay-back period. If the IRR is greater than 0, then the project is viable. (Most spreadsheet packages contain a series of macros or utilities for calculating IRR.) If the IRR is greater than the MCC for the company, then it is profitable.
- *Pay-back period* involves calculating the period (year) by which the accumulated Present Value of Returns equals the sum of the Present Value of Costs.

Caution: At the early phases of a project, such as during the Feasibility Study and Requirements Analysis, all the figures are estimates and no more. So the team should subject their calculations to a Sensitivity Analysis (recall Chapter 3). What this means, simply, is to try to range the calculations based on three sets of costs and figures: optimistic, realistic, and worst case. As an example, the project may have a marginal ROI based on optimistic figures. On realistic figures, the ROI is negative and on worst-case ranges, the ROI is a large negative. In this case, the sponsors may think twice about the project and either stop it, reduce their requirements, or do it for other nonfinancial reasons.

As the project proceeds, there will be numerous project planning sessions. At each session, the benefits and costs (both actual and estimated) should be evaluated, refined, and revised as shown in Figure 7.2. Should there be any major variance in the returns and costs equation, the project manager and team must subject these to the formal Change Control mechanism described in Chapter 4.

Actual costs would be collected during project tracking sessions and summarized for comparison with the initial Return-on-Investment Analysis during the project review session discussed in Chapter 4.

Again, the role of return-on-investment analysis in contemporary project management is essential in ensuring that the business perspective of projects is managed effectively. Projects are an investment by the project sponsor, and it is the project manager's responsibility to review and manage the ROI as rigorously as the other issues covered in this handbook.

The return-on-investment of a project cannot be altered without the project sponsor's approval.

Putting It All Together—Costwise

Figure 7.6 shows an example of an integrated Return on Investment model. It also includes an example of alternative solutions costing. The returns (financial and nonfinancial) and the costs have been calculated for each effective physical/design option. Both costs and returns have been extrapolated over the expected life of the system. The example includes Net Present Value, Payback Period and ROI calculations.

In Figure 7.6, the system will take one year to develop, and the returns do not begin accruing until the system is in production. Option A uses a manual implementation strategy with a recurrent people cost. Option B uses a semiautomated (computer/person) implementation strategy with a higher development investment but a lower recurrent cost per year. Both options have a positive ROI and, all other things being equal, Option A has a stronger financial ROI case.

Of course, all things usually are not equal, and additional nonfinancial returns may be associated with each physical implementation strategy to consider as well. As a refresher from earlier in this chapter, these will be in the Improve Service category.

For example, with Option B, the use of a computer will give the company a high-tech profile to match that of their competitors, and other people can use the computer for other purposes in slack periods. Assuming there are no strong additional nonfinancial benefits associated with the manual strategy (Option A), Option B may be more appropriate.

Mary Parker, Robert Benson and Ed Trainer [1988] argue that traditional ROI analysis does not take into account such factors as competitive advantage and technological infrastructure when analyzing the justification of a project. Should the organization wish to examine additional factors beyond those covered in this chapter, the Parker, Benson and Trainer book provides a simple five point ranking scheme for including additional factors.

Now for Something Completely Different

One other approach to Cost/Benefit analysis should be quickly discussed. The Cost Effectiveness approach ignores returns and simply uses a projection of the cost of maintaining the current system at the status quo compared with the cost of developing and supporting the proposed alternative new system. This ap-

STEP 1 - CALCULATE FIGURES IN CURRENT DOLLARS

INVOICE PROJECT - AARDVARK IMPORTS PTY LTD

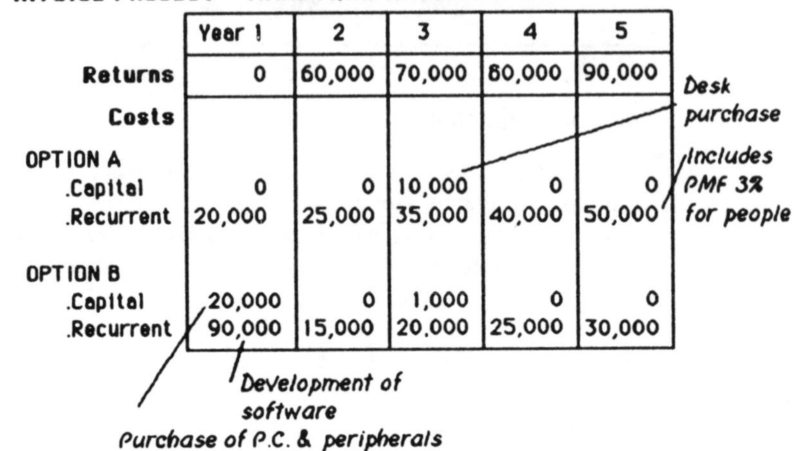

	Year 1	2	3	4	5
Returns	0	60,000	70,000	80,000	90,000
Costs					
OPTION A					
.Capital	0	0	10,000	0	0
.Recurrent	20,000	25,000	35,000	40,000	50,000
OPTION B					
.Capital	20,000	0	1,000	0	0
.Recurrent	90,000	15,000	20,000	25,000	30,000

STEP 2 - CONVERT TO PRESENT VALUE

INVOICE PROJECT - AARDVARK IMPORTS PTY LTD

	Year 1	2	3	4	5	TOTAL
Returns	0	44,592	44,849	44,184	44,849	176,474
Costs						
OPTION A						
.Capital	0	0	6,407	0	0	
.Recurrent	17,242	18,580	22,425	22,092	23,805	110,551
OPTION B						
.Capital	17,242	0	641	0	0	
.Recurrent	77,589	11,148	12,814	13,808	14,283	147,525

Discount Rate 15% p.a.

STEP 3 - CALCULATE NET PRESENT VALUE/R.O.I/PAYBACK

OPTION A - Net Present Value = \$176.474 - \$110,551= \$65,923

R.O.I. $= \frac{\$65,923}{\$110,551} * 100\% = 60\%$

Payback Period = 2.47 years

OPTION B - Net Present Value = \$176,474 - \$147,525 = \$28,949

R.O.I. $= \frac{\$28,949}{\$147,525} * 100\% = 20\%$

Payback Period = 3.52 years

Fig. 7.6
Putting It Together Costwise

proach says that if the cost of building and maintaining a new system is cheaper than maintaining the present system, then it is worth doing it.

This technique is really only appropriate if the team is not changing the functional requirements of the current system, as any additions to the status quo offered by the new system would also have to be ''added back'' to the status quo (which is no longer the status quo). However, this approach can be used to rate two alternatives that meet the same objectives or for conversion projects.

References

Commonwealth Department of Finance, *Guidelines for Costing of Government Activities.* A.G.P.S., Canberra, 1991.

P. F. Drucker, ''The Entrepreneurial Mystique,'' Inc. October 1985.

C. Gane & T. Sarson, *Structured Systems Analysis: Tools and Techniques.* Englewood Cliffs, N.J.: Prentice-Hall, 1979.

T. Gilb, *Principles of Software Engineering Management.* Reading, Mass.: Addison-Wesley, 1988.

M. M. Parker, R. J. Benson & H. E. Trainor, *Information Economics.* Englewood Cliffs, N.J.: Prentice Hall, 1988.

Risk Assessment Model

Don't worry, I'll think of Something
Indiana Jones, **Raiders of the Lost Ark**

Introduction

Risk assessment is the formalization of an "intuitive" process that has always been undertaken by project managers when planning a project. The risk assessment model contained in this Chapter is intended to introduce some rigor, objectivity, and consistency into what is typically a subjective and covert process.

Risk is critical in understanding the dynamics of projects. As shown in Figure 8.1, project risk has an impact on four key project management concerns.

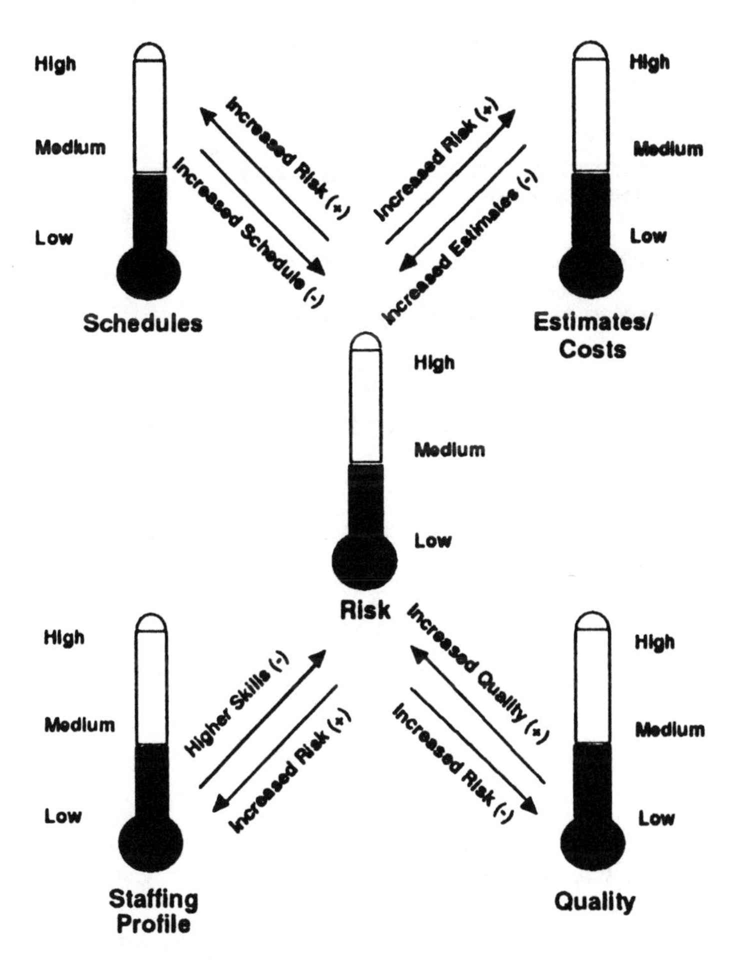

Fig. 8.1
Risk Dynamics

Unresolved Risk Factor Memorandum
Project : *Financial Management*

Risk Factor

Dependence upon vendor to deliver new release of software

Risk Factor Impact

. ***Project will be delayed for each day the new release is not available and tested***
. ***If new version not delivered by 8 July then Return-on-investment impacted negatively***

Risk Minimisation Strategies

. ***Obtain legal contract from vendor with penalties***
. ***Meet with vendor developers monthly***
. ***Add bonus to vendor contract for early delivery***

Contigency Plans

Prepare an alternative interim solution for system using earlier version of software

Fig. 8.2
Risk Memorandum

In general, the higher the risk of a project, the lower the expected quality, the higher the estimates/costs, and the longer the schedule. Conversely:

- The higher the required quality for the project's deliverables, the higher the risk.
- The shorter the schedules for the project, the higher the risk.

- The more severe the cost constraints, the higher the risk.
- The higher the skills profile of the team, the lower the project risk.

As discussed in Chapter 3, risk assessment is undertaken during the Rapid Application Planning sessions. Generally, risk assessment is done in conjunction with the evaluation and selection of the project development strategy and prior to estimation. Any risk factors ranked as high risk should be subjected to full analysis to determine actions required to reduce or constrain the risk *before* the project commences.

The project manager and team must seek the assistance of senior management, project sponsors and stakeholders in proactive reduction of risk. Reduction of risk in a project is a ''win-win'' situation with all project people gaining from the increased possibility of project success. It is typical, however, for a project to be exposed to risks that are beyond the project manager and his or her organization to resolve. Dependence on outside organizations is one such risk factor.

All high risk factors that cannot be constrained or eliminated during the RAP sessions should have a Risk Memorandum developed for them documenting the risk, the impact of the risk on the project, what actions can the Steering Committee and project sponsor take to assist in reducing the risk, and, for high impact risks, a contingency plan. Figure 8.2 shows a sample form for this purpose.

Risk can change as the project progresses. For example, it is possible for a project initially assessed as low risk to quickly escalate into a high risk project. Any alteration of project risk factors must be subject to the change-control mechanism detailed in Chapter 4.

The Questionnaire

Three main criteria have been identified as contributing to the risk of a project: the user environment, the team environment, and the system complexity. Each of these criteria has a number of related factors that have been allocated a low, medium, or high score. The sum of the factor scores provides an indication of the degree of risk in the total project.

It is also essential to understand that, as risk assessment is subjective, different people will perceive risks differently. Risk assessment should record all views democratically with the majority view being accepted as the guide. Should a split decision result from the democratic process, then the higher risk factor should be used—it pays to be paranoid!

A guide to the value for each factor is shown against each question but any value within the specified range may be used.

Risk Assessment Weights

The results of the questionnaire should be compared against the following table:

	Low Risk	*Medium Risk*	*High Risk*
User environment	32–127	128–223	224–330
Team environment	32–110	111–190	200–269
System complexity	33–126	127–218	219–308
Total Project Risk	96–363	366–631	633–907

Given the subjective nature of the process, the use of scores to compare projects is not valid. For example, a project with a total risk score of 650 is still a high risk project, as is one with a total risk score of 895. The use of the categories of low, medium, and high is the acceptable approach.

Risk Assessment Questionnaire

Risk Criterion: User Environment

	Weighted Risk Factor	Selected Weight
1. Will user commit to a standardized development methodology?		
Yes	Low	0
No	High	4
2. Has user been briefed and is prepared to commit to change control procedures?		
Yes	Low	0
No	High	4
3. How committed is senior user management to system?		
Extremely enthusiastic	Low	2
Supportive	Medium	6
Neutral	High	8
4. Priority for project within user area?		
High	Low	2
Low	Medium	4
Varied—more than one branch with different priority	High	8
5. How critical will system be to the user area continuing operations when completed?		
Minor impact	Low	2

Significant impact	Medium	4
Major impact	High	12

6. Number of outside organizations involved in concurrence, approvals, and other decisions relating to the system?

None	Low	0
1	High	16
Greater than 1	Very high	20

7. Number of different user branches involved in concurrence, approvals, and other decisions relating to the system?

1	Low	4
2	Medium	12
Greater than 2	High	20

8. Number of user sections or subbranches involved in concurrence approvals and other decisions relating to the system?

1	Low	1
2	Medium	5
Greater than 2	High	10

9. Number of unions involved in the system?

None	Low	0
1	Medium	10
Greater than 1	High	20

10. Number of user sites/installations involved with the system?

1	Low	2
2–10	Medium	6
Greater than 10	High	12

11. What is the severity of procedural changes/disruption in user department caused by proposed system?

Minor	Low	0
Significant	Medium	8
Major	High	12

12. Does user organization have to change structurally for proposed system?

Minor	Low	0
Significant	Medium	8
Major	High	16

13. Stability of system requirements?

Requirements are stable and unlikely to change	Low	2

Requirements are stable but likely to change	Medium	10
Requirements are unstable and open to further change	High	20

14. Status of current system documentation?

Not applicable	Low	0
Complete (may need rework)	Low	2
Acceptable but incomplete	Medium	4
None available	High	6

15. What percentage of existing functions are to be replaced on a one-to-one basis?

0%	Low	0
Less than 25%	Low	2
25–50%	Medium	4
50–100%	High	6

16. Available prototype or model?

Similar system in existence	Low	2
Subfunctions exist	Medium	4
Subfunctions exist across branches	Medium	6
Nothing done like this before	High	8

17. Project deadlines are?

Flexible—may be established in conjunction with project team	Low	1
Firm—established internally but missed dates may impact user functions/operations	Medium	6
Fixed—established by specific operations, legal requirements, direction beyond organization's control	High	16

18. User particpation?

Fully committed—expert user allocated to undertake significant project work	Low	4
Significant responsibilities—not full-time commitment	Medium	12
Some responsibilities—limited to review and approvals	High	16

19. How knowledge is user representative in proposed application area?

Understands area and involved in previous implementation	Low	2
Some experience	Medium	8
Limited	High	20

20. How knowledgeable is user in information systems development?

High degree of capability	Low	3
Previous exposure but limited knowledge	Medium	6
First exposure	High	12

21. What are communications between user area and IS like?

Good	Low	3
Fair	Medium	6
Poor	High	9

22. Is new user-controlled technology/techniques (e.g., monitoring equipment, graphics terminals, cad/cam, etc.) required for the system?

No	Low	0
Yes	High	9

23. Is the project dependent on a single user expert?

No	Low	0
Yes	High	20

24. Is there government legislation that the project is dependent on to meet deadlines?

No	Low	0
Yes	High	20

25. Is the project dependent on vendors and outside consultants/experts to meet deadlines?

No	Low	0
Outside experts	Medium	8
Vendors	High	12
Vendors and outside experts	Very High	20

Risk Criterion: Team Environment

1. Priority of project within IS is?

High	Low	2
Medium	Medium	4
Low	High	8

2. How committed is senior IS management to system?

Enthusiastic	Low	2
Supportive	Medium	6
Neutral	High	8

3. Project team size (including full-time user professionals)?

Less than 5	Low	4
5–10	Medium	8
Over 10	High	16

4. Total estimated development time in person months for system?

Less than 3 months	Low	4
3–20 months	Medium	8
Over 20 months	High	16

5. Estimated project development time?

Less than 3 months (elapsed)	Low	4
3–6 months (elapsed)	Medium	8
Over 6 months (elapsed)	High	20

6. What is general attitude of user area towards computing?

Good—understands value of IS	Low	3
Fair—some reluctance	Medium	6
Poor—anti-information system solution	High	12

7. How knowledge is user in area of IS?

High degree of capability	Low	3
Previous exposure but limited experience	Medium	6
First exposure	High	12

8. Project Manager (PM) availability, experience and training?

PM with successful recent experience in managing a similar project (type and size)	Low	4
PM with successful recent experience in managing either part of a similar project or other projects	Medium	6
PM with knowledge but little project managerial experience	Medium	8
Inexperienced PM	High	16

9. Key project skill and staffing level requirements can be met by?

Team members full-time	Low	3
Mix of full-time team members and part-time outside specialists	Medium	6
Part-time team members and outside specialists	High	9

10. What proportion of project team will be brought in from an outside company?

None	Low	0
Less than 25%	Medium	8
Over 25%	High	16

11. Number of team members who have worked successfully together on previous projects?

One person team	Low	0
All	Low	3
Some	Medium	6
None	High	12

12. How knowledgeable is IS project team in proposed application area?

Has been involved in prior implementation efforts	Low	2
Understands application area but no implementation experience	Medium	8
Mixed	Medium	6
Limited	High	12

13. What experience does the IS project team have with the programming language/s to be used?

Significant within the organization	Low	0
Significant in other organization	Medium	8
Little experience	High	16

14. What experience does the IS project team have with the data base system to be used?

No data base to be used	Low	0
Significant within the organization	Low	0
Significant in other organization	Medium	8
Little experience	High	16

15. What experience does the IS project team have with the data communications?

Data communications not to be used	Low	0
Significant within the organization	Low	0
Significant in other organization	Medium	8
Little experience	High	16

16. What experience does the IS project team have with the packages to be used?

Packages not to be used	Low	0
Significant within the organization	Low	0

Significant in other organization	Medium	8
Little experience	High	16

17. New operations system installation required?

No	Low	0
Yes	High	12

18. Which of the hardware is new technology to the project team?

None or	Low	0
CPU and/or	High	12
Peripheral and/or	Medium	6
Telecom (e.g., remote lines)	High	12
Terminals and/or	Medium	6
Other (e.g., mini, micro, etc.)	High	12

Risk Criterion: System Complexity

1. Proposal document from users/analyst?

Complete	Low	0
Acceptable but incomplete	Medium	2
None	High	4

2. Expected operational life of system?

One-off	Low	1
Short life (<1 year operation)	Medium	4
On-going function	High	12

3. System performance is?

Not critical or/	Low	0
Critical due to throughput/volume and/or	High	16
Critical due to space/ memory and/or	High	16
Critical due to response time	High	16

4. Current documentation for existing computer systems?

Complete (may need rework)	Low	3

Acceptable but incomplete	Medium	6
None available	High	13

5. What percentage of existing functions are to be replaced on a one-to-one basis?

0	Low	0
Less than 25%	Low	4
24–50%	Medium	8
50–100%	High	12

6. Available prototype or model?

Similar system in existence	Low	1
Subfunctions exist	Medium	4
Subfunctions exist across branches	Medium	6
None available	High	12

7. The software will be?

Hardware independent	Low	0
Dependent on parts of hardware	Medium	6
Completely hardware dependent	High	16

8. Additional hardware resources of a type already used for project is?

Not required	Low	0
Required	High	4

9. Availability of required additional hardware resources?

Available	Low	0
Limited	Medium	5
None	High	16

10. Does the system have logical (user-driven) complexity as measured by: Logical input types (e.g., unique input screens, records, transaction?)

1–5	Low	3
5–40	Medium	6
40+	High	12

11. Does the system have logical (user-driven) complexity as measured by: Logical output types (e.g., unique output screens, report layouts?)

1–5	Low	3
5–40	Medium	12
40+	High	20

12. Does the system have logical (user-driven) complexity as measured by: Logical user views of data (e.g. unique logical files, DB2 tables?)

1–5	Low	3
5–20	Med	12
20+	High	16

13. Does the system have logical (user-driven) complexity as measured by: Automated transfer of input/output files to other systems?

1–5	Low	3
5–20	Medium	12
20+	High	16

14. Does the system have logical (user-driven) complexity as measured by: Logical enquiry types?

1–5	Low	3
5–40	Medium	12
40+	High	20

15. Amount and quality of data to be transferred into the system on initial load as measured by logical user views?

1–5	Low	2
5–20	Medium	4
20+	Medium	8

16. Complexity of data element editing is?

Simple	Low	3
Complex	Medium	9
Highly Complex	High	12

17. Does data require viewing and/or update security?

No	Low	3
Yes	High	6

18. Complexity of algorithmic processing is?

Simple	Low	3
Complex	Medium	9
High complex	High	16

19. The software developed must interface with?

Stand alone	Low	0
Systems under teams control	Medium	3
Systems under others control	Medium	6
Complex systems under team control	High	10

	Complex systems under others control	High	16
20.	The software will consist of?		
	Total outside package that meets all system requirements and/or	Low	0
	Precoded in-house modules written by project team members and/or	Medium	6
	Precoded in-house modules written by other IS staff and/or	Medium	6
	Precoded/packaged modules written by outside vendors/ sources and/or	Medium	8
	Total outside package that meets 90 percent of system requirements and/or	High	10
	Only modules written for this system	High	10
20.	What level language will be used in the system?		
	Very high (SQL, Telon)	Low	2
	High (PL/1, Fortran, COBOL)	Low	6
	Medium (basic, CSP)	Medium	10
	Low (Assembler, CICS)	High	20

Software Metrics Revisited

As discussed in Chapter 3, the quantification of risk factors is termed Software Metrics. As much as possible, the Risk Model contained in this chapter provides measures to assist in determining the scale or impact of the risk factors. However, many of he risk factors remain subjective. For example, there are a few practical ways of measuring and enthusiasm level of businesspeople toward a project. As a result, a full risk model remains a mix of both objective and subjective measures.

Whenever possible, the project manager should refer to the refernce sources on Software Metrics such as those by Capers Jones [op. cit.] and others to support their subjective assessment of risk. In addition, the full involvement of team members and project stakeholders in assessing risk provides an invaluable mechanism for putting assumptions and differing views on the table. As for many of the techniques in this handbook, the very process of undertaking risk assessment is valuable in sharing assumptions and concerns, and in building a common vision for the project.

Different Risk Models

It is important to distinguish between the risks inherent in the *process* of developing the product and the risks associated with project failure. The risk model in the chapter is a process or product risk model. Some organizations also attempt to predict the risks associated with the impact of project failure, that is, non-delivery, cost overruns, etc.

For example, at a simple level, the risks associated with project failure may be simply financial—the cost of the project must be written off. In more complex projects, the risks may be:

- *Political*: The organization is in conflict with government legislation.
- *Human resource*: Staff are disaffected.
- *Financial*: The organization loses money, capital equipment.
- *Security*: The organization is exposed to fraud, illegal access to data, and the like.
- *Legal*: The organization faces litigation.
- *Market*: The organization loses image or market share.

These risks are identified during the project planning and are normally added to the Business Case for consideration during project approval.

In summary, project development risks models (as in this chapter) provide an indication of the probability of project failure and subsequent exposure to these broader risks.

Advanced Issues and Hints

Any good idea tends to institutionalize itself.
Curry's Caveat [1973]

This chapter is a potpourri of issues that are additional components involved in the Thomsett Associates' approach to project management. The following issues are covered:

- How to conduct a Rapid Application Planning session.
- Modeling scope.
- The role of quality assurance.
- Quality requirements and estimates.
- Wide-band Delphi estimation.
- Some rough rules of project management.

How to Conduct a RAP Session

The processes involved in a team-based planning session have been covered in Chapters 2 and 3. The following points deal with the environmental and people aspects of the planning session.

The concept of planning sessions, or Rapid Applications Planning sessions, is similar to the concept of Joint Application Design sessions used in the system development cycle. That is, by focusing a key group in an intensive effort, the quality of output is increased and the costs and elapsed time are reduced.

Firstly, the room or space in which the planning session is to be conducted should be equipped with at least two white-boards, preferably the electronic types that can produce copies of what is on the board. Second, there should be adequate space to enable the teams to break into smaller subteams when they are listing tasks and estimating tasks. Third, there should be adequate big sheets of paper for publicly recording issues as they arise for later discussion. Finally, all public—that is, large group discussions—should be tape-recorded or at least summarized at regular (½- to 1-hour) intervals.

The conduct of the planning session should be fairly disciplined. That is, the planning session follows each step of the planning process outlined in Chapter 3 in sequence (Review Business Case, Select Strategy, Risk Assessment, and so on). The chairperson (who is typically the project manager) should ensure that the planning session focuses on project management issues, *not* on technical issues; any technical issues should be recorded on one of those large sheets of paper for later technical sessions.

As in many meetings, the project planning sessions may tend to be dominated by a few people. This will be common since as many of the team members will not have participated in planning sessions before or may see themselves as purely technical people who should not be involved in the planning process. Whenever possible the meeting chairperson should attempt to ensure that all people at the planning session are involved by breaking the team into smaller subteams at various stages of the planning session. As already mentioned, listing project tasks, estimating tasks, and risk assessment are all project planning activities that can be used for small team work, with the various small team outputs being combined and merged into a single team output.

If Level 1–3 groups or observers are involved, try the "fish-bowl" meeting technique, by which key people participate in the center and the observers watch from an outer area.

For larger projects, the planning sessions may take a couple of days. In this case, the use of half-day planning sessions (with the other half-day left to gather additional material or for other work) can often help the flow of creative input. In any case, project planning sessions tend to be fairly intensive; so remember to include lots of breaks and coffee between sessions.

In some cases, the use of a neutral facilitator, such as a person from the Quality Assurance or Development Support groups or from a Staff Development area, can be helpful in large planning sessions.

Modeling Scope

One of the more challenging areas of project planning is the question of project scope.

To many people, scope and objectives are the same. It is probably more useful to view scope and objectives as interrelated but different. The *scope* of a project defines the boundaries of the project manager's responsibilities, and the *objectives* of the project define what the project manager must achieve within those boundaries.

In information system projects, one of the more critical issues is the difference, if any, between *information system* scope and *project* scope (see Figure 9.1). This is especially critical in the case of large systems (see Chapters 10 and 11), where the noncomputer systems elements may be significantly larger than the computer systems elements. Typically, the smaller the project, the more likely that the system and project scope are similar.

The scope of an information system can be defined graphically using contemporary technical specification techniques such as data flow diagrams and data models (Figure 9.2). However, as mentioned earlier except for small projects, the information system scope is a subset of the project scope.

All information systems impact the organization and its people in some manner. Often, the successful implementation of a new information system or enhancement will require related job redesign, physical office alteration, new procedures, and new managerial control patterns. In addition, large projects may involve legal issues, financial management, and extensive administrative support such as travel and accommodation.

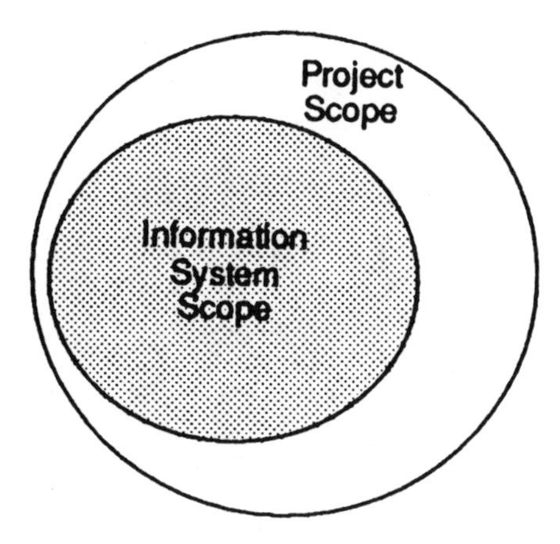

Fig. 9.1
Project and System Scope

These projects require a number of noninformation system activities (often undertaken by business experts).

It is essential that the project scope and objectives reflect the noninformation system objectives as well as the information system objectives. Equally important is the understanding of the risks, tasks, estimates, and quality requirements of the nonsystem components of the project. The involvement of Level 3 and higher level stakeholders will usually ensure that these issues are covered in the RAP session. In many projects in which Tomsett Associates has been involved, the nonsystem components of a project have been on the critical path and have been the highest risk areas.

One very useful tip for stating project scope is the modified use of a technique developed by Kepner and Tregoe [1981] for problem statement. As shown in Figure 9.3, this technique involves stating what the scope "is," what it "is not," and (in early stages of the project planning) what "is not known." Of course, any "not known" issues need to be resolved as a matter of urgency.

The Role of Quality Assurance

Quality is generally perceived as a technical issue. However, it has a significant impact on project management. It is the nature of projects that they consist of a series of tasks that "add value" by processing the project's product through a series of dependent tasks. For example, the outputs from the Requirements Analysis phase are passed into the System Design phase.

As shown in Figure 9.5, the dependent nature of project tasks means that any defect (human or machine-induced error) in one task is passed on to the next

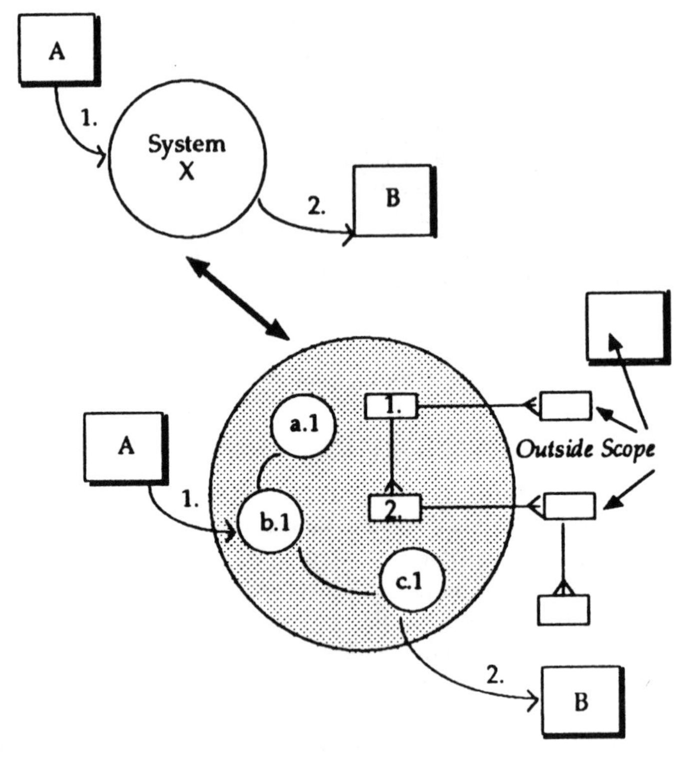

Fig. 9.2
Information System Scope

task, where it may be detected and add unscheduled and unplanned time and effort in removing the defect.

In other words, Task A passes input to Task B. Task B was scheduled, for example, as 4 days of work effort and 8 elapsed days duration (recall Chapter 4). However, some of the defects from Task A are detected during Task B, and these defects require another 1 day of effort. Task B now has 5 days of work effort to allocate but still only 8 days of duration. In other words, the person undertaking Task B appears to have misestimated the actual effort for Task B. In some cases, if person B is not prepared to work harder, the elapsed duration for B will "blow out" to 10 days. Of course this process continues in a more significant way for Task C, which has to account for defects from Task A and B.

The impact of this in a project of 80+ tasks is clearly massive and the traditional technique of phase-end reviews will not be able to detect all the accumulated defects.

This "defect-ripple" effect has been recorded by Capers Jones [op. cit.]

Scope Statement Project : *Smarm Office Automation*		
Is	**Is Not**	**Not Known**
Installation of work-stations	*Supply of LAN's*	*Definition of Common User Interface*
Design of training	*Delivery of training*	
Production of user guides	*Support of software*	
Selection of software		

Fig. 9.4
Modified Kepner-Tregoe Scope Model

and others as accounting for up to 40 percent of the total effort of a project over its development cycle. In other words, without an effective quality assurance process, a major cause of the difference between estimated effort and actual effort is the "defect ripple" effect.

Simply, poor quality assurance is one of the biggest causes of poor estimation and project slippage. It is obvious that, by the middle of a project, the actual effort required to remove embedded defects passed on from all the previous tasks can exceed the estimated effort required to perform the task currently being completed.

It is agreed by all experts on quality that the most effective way of detecting and removing defects is to quality assure and review *each* task's output *before* it is passed on to the next dependent task. As a rough rule of thumb, each task should be expanded by 10 percent to allow for the quality assurance process to

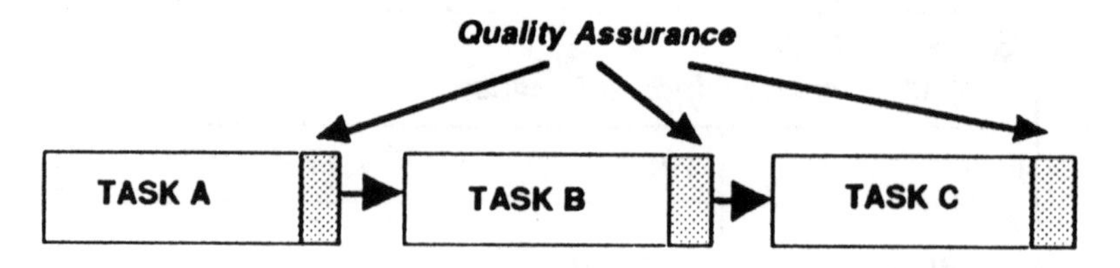

Fig. 9.5
Defect Ripple

occur before the commencement of the next task (see Figure 9.5). It is also essential that the Project Tracking documents (time recording) have a recording category for defect repair and removal.

> **Do not despair if the amount of time recorded for defect repair is high, as it is better to find them than to hide them.**

Quality Requirements and Estimates

Another important aspect of quality in relation to project management is the definition of quality with respect to requirements and the impact of quality on estimates.

Quality Requirements

In the case of computer software, it is generally agreed that, based on the work of McCall [1980] and others, quality consists of the following attributes:

- *Conformity*: Does the product have all the data and function specified?

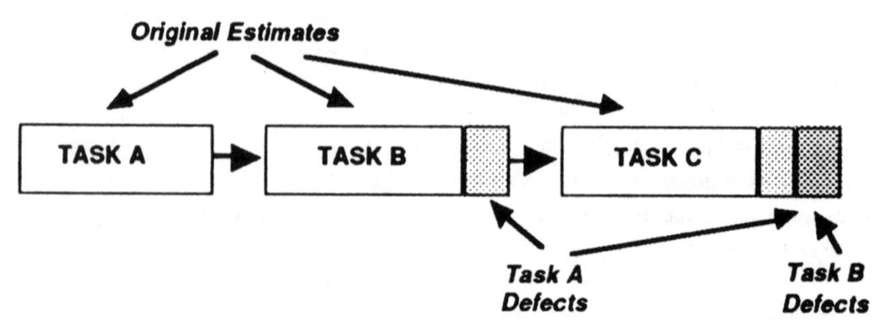

Fig. 9.6
Allowing for Quality

- *Usability*: Is the software easy to use and understand from the client's perspective?
- *Efficiency*: Does the software use the hardware, data base, and other support software efficiently?
- *Maintainability*: Is the software easy to maintain and support?
- *Response time*: Does the software provide an adequate response time for the clients?
- *Flexibility*: Is the software easy to modify to include new function and data?
- *Portability*: Can the software easily operate in different software and hardware environments?
- *Security*: Is the software secure from unauthorized access and modification?
- *Auditability*: Can the software be easily audited and does it include adequate controls?
- *Job impact*: Does the software affect the existing work flows, control, and autonomy of the user area?

Each of these quality attributes has a subset of criteria. For example, conformity is a measure of completeness (is all the data and function present?) correctness (is the data and function defect free?) and traceability (can the data and function be tracked from requirements, through design and into production?). The relationship between Quality Attributes and Quality Criteria is shown in detail in Appendix B.

It is clear that some of these attributes are potentially incompatible. For example, security could degrade response time. Usability could degrade efficiency. Further, many realists add meeting deadlines as another quality attribute especially if quality includes senior management concerns. Deadlines are clearly incompatible with all the other quality attributes. Many projects meet deadlines, which is a mandatory quality requirement, by degrading conformity, usability, auditablity, and so on.

A wise and prudent project manager should attempt to gain a consensus on what quality attributes are required for the project's deliverables *before* planning the project and negotiate a formal Quality Agreement (see Appendix B).

This is essential where the project has a number of Level 4, 5, and 6 external groups as covered in Chapter 5. The project manager and team should get each external group to specify which quality attributes are Mandatory and which are If Possible (or using the Kepner-Tregoe approach, which criteria are Musts and which are Shoulds). This process produces a Software Quality Agreement.

Ideally, the project team would undertake some scoring of the quality criteria for each of the mandatory quality attributes. Unfortunately, many of the software quality criteria such as Useability are difficult and expensive to measure; so a more pragmatic approach is to use a subjective Group Nominal technique of scoring the criteria within a range from the worst to the best. Figure 9.7 shows an example of this technique, and Appendix B has a complete model for use in defining and scoring the project's software quality requirements.

Typically different external groups perceive the required quality attributes differently. In this case, the project manager should raise the potential conflicts with the Steering Committee for resolution.

Finally, as discussed in Chapters 4 and 5, during the project development cycle, various changes to the quality requirements will occur and these should be subject to a formal change-control process.

Changes in quality attributes are a common but a subtle form of change, which should be monitored and controlled as for any other change.

Quality and estimates

In the late 1960s, Gerry Weinberg [1971] conducted an experiment with a number of computing teams. Given the same data and functional requirements but differing quality requirements, one team was asked to produce the most readable output while another team was asked to minimize the use of CPU. Each team, while meeting its quality requirement, required very different effort to achieve this. The productivity variance was 7:1!

Simply, different quality attributes have a different impact on the project estimates. While some attributes such as maintainability and flexibility have relatively low impact—that is, a requirement to make a system maintainable would not add a significant cost to the project—others have a major impact. In particular, high requirements for job impact, efficiency, usability, and portability require significant effort to achieve.

It is essential that the project manager and team clearly understand the requisite quality requirements *before* they finalize the project estimates.

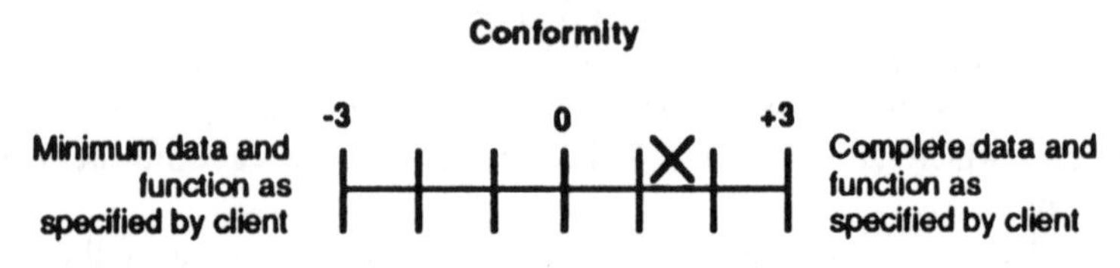

Fig. 9.7
Scoring Software Quality Criteria

Wide-Band Delphi Estimation

In the absence of any formal estimating history, estimation in many projects would be better termed "guestimation." In some texts, guestimation is defined as a guess based on internalized experience or "gut-feel."

The following technique for guestimation based on the Delphi technique developed by Herman Kahn of the Hudson Institute has been shown to be very effective in many projects. It is a team-based technique that is easily embedded in the team-based project planning approach advocated in this handbook. It involves nine simple steps:

- *Step 1*: Provide team members with the relevant information regarding the project, that is, the Business Case, quality requirements.
- *Step 2*: Conduct a formal Risk Assessment as described in Chapters 3 and 9.
- *Step 3*: Develop task lists as described in Chapter 3.
- *Step 4*: Each person individually estimates each task using Sensitivity Analysis to provide a best case, likely, and worst case estimate (see Chapter 3).
- *Step 5*: All estimates are written on to a white-board and grouped in the three ranges.
- *Step 6*: Each person discusses the various assumptions and issues they considered when developing their estimates.
- *Step 7*: Where required, the various estimates are adjusted based on the team discussion.
- *Step 8*: Each range is averaged with outriders being discarded.
- *Step 9*: The resultant ranges are used as the basis for scheduling as discussed in Chapter 3.

This results in a highly discussed ranged set of estimates as shown in Figure 9.8.

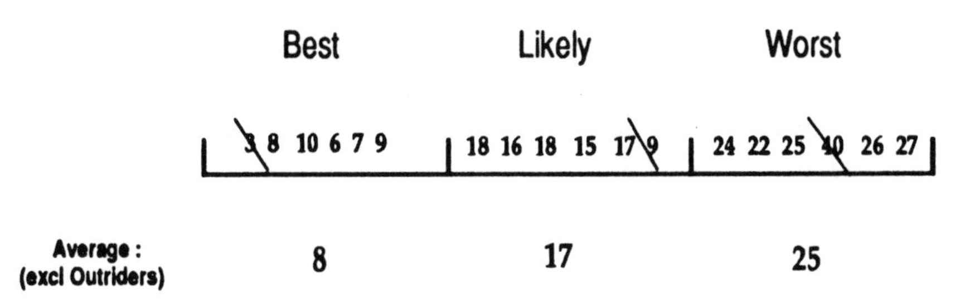

Fig. 9.8
Wide-Band Delphi Estimates

This variation on the Delphi technique will work very well, provided that a free-ranging discussion about each participant's assumptions while deriving their estimates (Step 6) is allowed and that the original estimates from Step 5 are adjusted to reflect the discussions. In Figure 9.8, the outriders in each range have been deleted after discussion and before averaging the remaining estimates.

Again, it is the discussion incorporated in the Wide-Band Delphi technique that is important. The team and stakeholders learn more about their various assumptions during the estimation process.

Some Rough Rules of Project Management

These rules summarize bitter lessons learned by many battle-hardened veterans:

- What you measure gets optimized.
  Any measure that is used to monitor people will be distorted to ensure that everyone is performing equally. For example, one team spends 5 work hours of effort per day and another spends only 3. As soon as the second team is made aware that they are not as effective as the other team they will immediately distort their recording of effort to equal 5 work hours of effort. Any measurement system must recognize the many factors that affect individual team productivity.
- In projects, single dimension measures of productivity and quality are always wrong.
  There are at least 100 factors which influence project quality and productivity —see Risk Assessment.
- You know when you have a working project management system when you do not need people's names.
  Project tracking tracks projects, not people.
- Most projects that fail had failed before they started.
  These projects were given fixed deadlines before any estimates, requirements, quality requirements, and resources were determined and so were never planned properly.
- It is extremely difficult for a Project Manager to stop a project that is failing.
  A project that is failing will become obsessed with working harder. Planning and quality begin to be seen as luxuries that the project cannot afford.
- It is extremely difficult for a Project Manager to stop a project that is succeeding.
  The reverse of the preceding rule is that a project that is succeeding can become too confident and begin to become too focused on delivery.
- There is no such thing as a simple project.
  In simple projects, people can become careless and use informal and/or verbal planning, which leads to misunderstanding and . . .

- Most simple projects degenerate quickly into less than simple projects because they are simple.
  See the preceding rule.
- If you cannot stop a project, it is failing.
  No matter how close the deadline, planning will always pay back the time required.
- Planning is work; working is not planning.
  If you are not sure of this, read this handbook again!
- In most projects, the team will know the true situation before the Project Manager.
  The team undertakes the technical work, and there are more of team members than there are project managers.
- It pays to be paranoid when planning a project.
  See Risk Assessment.
- A well-managed project is generally less exciting than a badly-managed project.
  The adrenaline rush of a project forced to meet a deadline or to solve a difficult problem under pressure is the stuff of heros and heroines.
- We should allow time to learn from every project—good or bad.
  There is nothing smart in thinking that your project is the only one that is going to face problems. The use of Post-Implementation Reviews is vital in this area.

References

T. C. Jones, *Programming Productivity: Issues for the Eighties*. Los Angeles: IEEE Computer Society, EHO186-7, 1981.

C. H. Kepner & B. B. Tregoe, *The New Rational Manager*. Princeton, N.J.: Princeton Research, 1981.

J. McCall and M. Matsumoto, *Software Quality Metrics Enhancements,* 2nd ed. Rome Air Development Center, Griffiths Air Force Base, N.Y.: 1980.

T. L. Martin, Jr., *Malice in Blunderland*. New York: McGraw-Hill, 1973.

G. M. Weinberg, *The Psychology of Computer Programming*. New York: Van Nostrand Reinhold, 1971.

10 Super-Large Projects

Hubris, n. Insolent pride or security, overweening pride leading to retributive justice.

The approaches introduced in this handbook are suitable for all projects. However, as predicted by Nolan [1979] in his Stage Hypothesis model, many organizations are either planning to or have commenced to redevelop vital and major systems to take advantage of second-generation relational data base and communications technology. As a result, the unique nature of these projects has special significance to organizations through the 1990s.

There are a number of special issues relating to the project management of large and super-large projects.

Large Projects—a Definition

Three major criteria and two secondary criteria are involved in defining project size. The major criteria are application/software size and complexity, development effort, and development time-scale. The secondary criteria are technology risk and organization impact.

Application/Software Size

Application size can be defined using three related groups of software metric: (1) lines of delivered source code (LOC); (2) input, output,and file volumes; and (3) IBM's Function Point [Dreger, 1989].

Capers Jones [op. cit.] defines large projects as those projects that deliver a system of 64,000 LOC or greater. Barry Boehm [1981] defines large systems as 128,000 LOC or greater. Other sources define a large system as one with more than 40–50 logical inputs, 40–60 outputs, 25–30 inquiry screens, and 20 plus logical data base views (files). Function Point sizing, which is also based on inputs, outputs, and so on, would suggest that a large system produces around 1500 Function Points or greater.

Both Jones and Boehm agree that there is a special category of information system—the super-large system that produces 512,000 LOC or greater.

A large system (for example, one with more than 128,000 LOC or greater than 1400 Function Points) may also be considered as a super-large system should it require complex input, output, file considerations or require complex algorithmic processing.

Development Effort

Development effort is typically defined measured in people effort, capital equipment investment and organization on-costs (accommodation, consumables, etc.).

Boehm and Jones provide estimating heuristics that put the development effort for large systems at 50,000 work hours plus or approximately 40–50 person years (based on 5 work hours/day and 210 days/year). In addition, the actual development effort would require additional user professional effort not usually measured by computer-oriented project tracking systems. These people would be indirectly involved in education, physical office reorganization, testing, file conversions, and so on.

Assuming no direct capital investment for the project, the project also consumes computer time and resources for development design, coding, file creation, and so on. A reasonable rule of thumb could use the standard metric that the Information System budget is split roughly 50/50 between people and equipment, software and services. In this case, each dollar spent on people is matched by a dollar of computer/software expenditure. Of course, many large projects involve direct capital investment of many millions of dollars.

Using a generally accepted costing of $40.00 per hour for a typical computer project member, the development cost of a large project would exceed $2,000,000. In addition, using the general rule that people/salary costs are around 50 percent of a typical information systems group budget, another $2,000,000 could be added for capital investment and equipment costs. Finally, most large projects require substantial effort from various user professionals, which could, in the cases of high organizational impact, equal the direct project effort and costs (that is, $2,000,000).

In summary, development costs for large projects can be of the order of $6,000,000 and greater.

Development Time Scale

Given the 40–50 person year effort required for large projects to build the system, the minimum development time scale is around 12–18 months.

However, as indicated by Boehm [op. cit.], Putnam [1978], Brooks [op. cit.] and others, the scheduling of 50 person-years effort on a single product in a year can lead to inefficient staffing peaks. Therefore, as described later in System Development Strategies (see also Chapter 3), large projects are divided into subprojects. These subprojects are scheduled and staffed relatively independently, and as a result the development time scale is generally staggered over a longer period. Staggered development schedules over two years or greater is common for large projects.

In a period of unprecedented economic, political, and technological turbulence, the longer the project exists over time, the higher the probability that the project will have to accommodate external requirements changes, technology changes, and internal changes such as staff attrition and turnover. In other words, the very requirement for long development time scales requires large projects to accommodate a high rate of change leading to a further expansion of the time scale.

Technology Risk

The use of leading-edge technology and/or stable technology to the limit of its processing capability is another feature of large projects. In many cases, technology risk is also sufficient for a medium system (64–128K LOC) to be treated as a large system from the management perspective and for a large system to be treated as a super-large system.

Current examples of technology risk would include the use of relational data bases, application generators, fourth-generation programming languages, and communications networks for large projects involving high data volumes or transaction rates.

The stretching of current technology would be typical for large projects. The size of these applications would require transaction rates, data base volumes, communications network, and systems design which are often beyond the design features and capability of current technology and as a result, introduce the exponential scale-of-effect multipliers.

This scale-of-effect was first documented by Brooks [op. cit.] in *The Mythical Man-Month,* in which he documents his learnings from managing the largest project of the time—the development of IBM's OS/360 operating system. This effect is shown in Figure 10.1.

Organizational Impact

The final criterion for large projects is the impact of the project on the organization's strategic growth path and the essential processing units in the

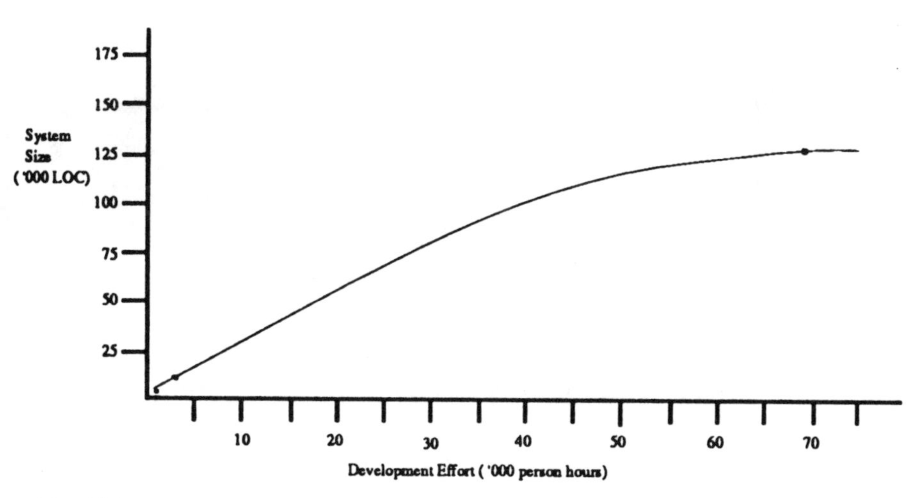

Fig. 10.1
Scale-of-effect Model (from Brooks, Jones)

organization. Typical large projects are vital to the survival of the organization and are a major element in the development of a new strategic product.

As a result, large projects require substantial reorganization and rationalization of many sections of the organization. This raises issues of job redesign, union involvement, client service facilities, and products, physical office redesign, and so on. Peter Keen [1981] is one of the few people who has documented the complex issues associated with the organizational change and its impact on people associated with large computing projects. In particular, he addresses the difficult political issues that will be raised by the high degree of change. It is common, as mentioned earlier, for higher costs and impact to be incurred outside the information system component than directly within the information system scope.

In addition, the development of large projects require a higher level of intraorganization coordination. For example, because of the wide organization impact, many separate user groups and technical support groups (data base, network engineering, operations, etc.) are involved in reviewing and approving planning, development, and implementation of the various new products and deliverables associated with the project. In many cases, outside organizations—such as vendors, contractors, unions, and other companies or government departments—are also involved with the project.

As detailed in in Figure 10.2, the three major criteria—application size, development effort, and development time scale—define large projects. Medium projects with leading edge technology and/or high organizational impact tend to behave as large projects and exhibit the same issues. For the purpose of this paper, they will be considered as large projects.

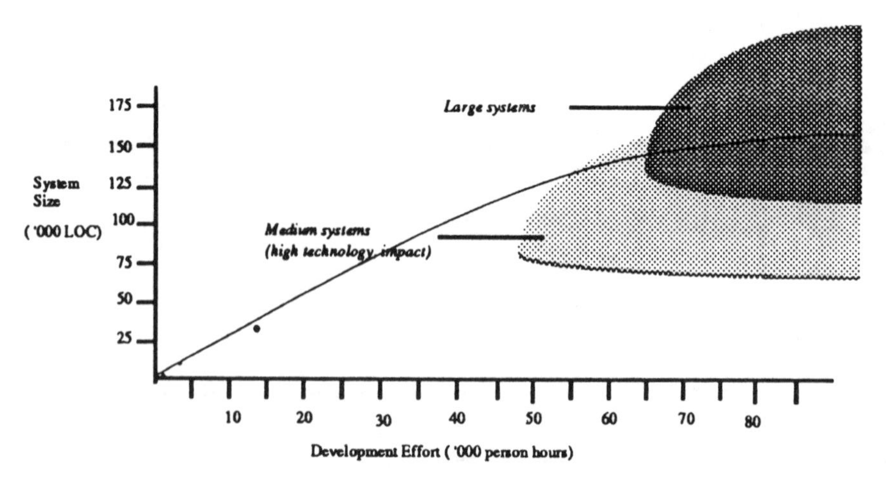

Fig. 10.2
Large Projects: Summary

Management Issues

The management of large projects centers around a set of concepts that are relevant to all projects but that require special focus and variation for large projects. These concepts are:

- Project success.
- Project management technique.
- Project team organization.
- System development strategies.
- Quality assurance.
- Support tools and technology.
- Stakeholder management.
- Team building and maintenance.
- Change-control procedures.
- Project agreements or contracts.

Project Success

It is generally accepted that a software project is successful when it: (a) meets agreed upon requirements; (b) meets agreed upon deadlines; and (c) meets budget as measured in people, equipment, and so on. In addition, it is reasonable to add a fourth success criterion: that the project team feels that it has done a professional job.

For large projects, the overwhelming evidence is that the concept of success must be reconsidered. As detailed in Selected Readings, every product development profession (engineering, construction, defense, manufacturing) has its own version of the large system dynamic. For example, currently in the Australian construction industry, the new Parliament House and the Darling Harbour project are clear examples of large projects.

In both cases, the publicly available material indicates substantial blow-out of costs (10:1 for the Parliament House and 2:1 for the Darling Harbour) and missing of deadlines and slippage of project time frames. Further, the Darling Harbour costs exclude the Casino which in January 1988 (the original deadline) had not even been started! While these may appear to be isolated cases, any detailed research into large manufacturing, defense, and construction projects reveals the same pattern of project cost-blowout and deadline slippage.

For example, the RAND Corporation [Stringer, 1982 (see Selected Readings)] examined the cost overruns of new Processing Plants in 1981. One of their conclusions was:

Severe underestimation of capital costs is the norm for all advanced technologies.

An examination of large computing projects within Australia and the United States reveals the same pattern. The Australian Public Service's MANDATA, Job-Seeker, and Stratplan projects, the Commonwealth Bank's Hogan project, Bank America's Masternet [Ludlam, 1988] and Allstate Insurance [Rothfeder, 1988] all failed to meet the three success criteria. In the case of MANDATA and the Hogan project, the projects were abandoned once it became clear that the requirements were facing compromise following budget and deadline blowout. Stratplan and Job-Seeker have both faced massive cost and schedule overruns but are still being developed since the requirements for the projects are still achievable.

The conclusion must be drawn that, for large projects, the inability to meet all three success criteria is not the exception *but the rule*. Simple, whereas normal projects could be expected to meet all three success criteria, large projects will meet only *one* of them—meet requirements *or* meet deadline *or* meet budget.

This revised concept of success for large and super-large projects does not imply that budgets and deadlines are simply sacrificed in order to meet requirements. Indeed, as argued later, the application of budget and schedule control is essential to minimize additional blowouts. However, for large projects, it is inevitable that meeting requirements (original and revised) will not be possible within the initial estimates and budget. When this occurs, the budget and schedule are expanded in a controlled process (see change-control later in this chapter).

Project Management Technique

Formal and highly disciplined project management techniques are mandatory for large projects.

Because of the need for communication across multiple subteams and specialist teams within and outside the project team's organization, the detailed scheduling of tasks and documentation of critical project management information—risks, assumptions, key decisions, agreements, and costs—is essential to provide a total project view.

Further, the use of organizational standards—such as development methodologies, naming conventions, data dictionaries, and formal Quality Assurance techniques—can minimize the communication complexity. For example, it is typical that large projects will involve multiple system development approaches such as information and functional modeling, prototyping, reusable code, application generators, and so on. These various methodologies should be formally documented and common among all subprojects.

The role of senior management is also vital in large projects. Whereas many Steering Committees simply act as review groups, because of the broad organizational impact of large projects, the Steering Committee must act as a problem-solving group particularly in the area of cross-boundary or intergroup disputes. Given the strategic importance of large projects, the senior management Steering Committees must become involved at a more detailed level than is typical for Steering Committees. As discussed later, changes are inevitable in large projects and changes to requirements, costs, and schedules must be reviewed by the Steering Committee, and specific levels of delegation and authority must be in place to ensure that the typical process of gradual and incremental alteration of the project does not occur without Steering Committee approval.

Thomsett Associates' research indicates that large projects involve additional project management effort than that required for normal projects. It is generally accepted that 10 percent of the total development effort should be allocated to the processes of project management for normal projects. For large projects, this allocation could rise to the range of 20–30 percent of the total development effort. In other words, on a large project of 50 people, 10 would be full-time on running the project management system—selecting tasks, estimating, scheduling, tracking effort, and deliverables, reviewing and rescheduling as required. Since much of the project management process involves clerical tasks such as recording, summarizing and aggregating data, and driving the automated project scheduling tools, the 10 people allocated to project management would include a number of clerical support people (similar to IBM's Programming Assistant concept).

Finally, given the large number of subprojects and related tasks requiring scheduling and tracking, the use of automated project scheduling tools is also essential for project management of large projects.

Project Team Organization

Clearly the scenario for large projects is one of a project management team with a number of subproject teams consisting of user professionals, computer professionals, clerical support, and computer specialists (data base, communications, etc.).

While each subproject may have its own project manager, the project management team is responsible for the total project. The central project management team could also include the project-resident technical specialists and the clerical support staff. In other words, the central project management team is the project's resource center for specialist functions. This is in addition to the normal Information System specialist areas.

Each subproject team is comprised of user and computer professionals working on the component of the overall system being developed by the subproject. As is now generally accepted, user professionals from the organization areas involved in the project are full-time members of the project team, and in many cases senior user professionals perform the project management functions.

It is mandatory in large projects to minimize the hierarchy and maximize the communication. While this may appear counterintuitive, the history of large projects (see Selected Readings) indicates a common pattern of many layers of management being involved in making even the most simple of project decisions. The result of this is massive delays in decision making, substantial filtering and distortion of project communication, and, in effect, collapse of the project management system for which the management hierarchy was formed to control and manage. In essence, the larger the project, the shorter the communication lines must be to ensure fast feedback through the project management system to all subproject teams.

Ideally, a large project should have no more than *two* levels of management between the subproject team and the Steering Committee. This can easily be accomplished through the use of Likert's [1971] linking-pin structure. As shown in Figure 10.3 this structure involves each subproject team being represented on the project management team and the project management team being directly involved in the Steering Committee team. In other words, the Steering Committee has the project manager as a full-time member and certain team members are involved in two teams (either horizontally or vertically) as the linking-pin.

Further, the use of an electronic mail system dedicated to the project team is a valuable aid in ensuring that critical messages are sent to all relevant people. Most electronic mail systems also include an ''action-prompt'' feature, which would ensure that critical messages were responded to within an agreed-upon time frame.

System Development Strategies

A system or project development strategy is the overall partitioning of the project and the high -level sequencing of subprojects. Paul Melichar [op. cit.] of

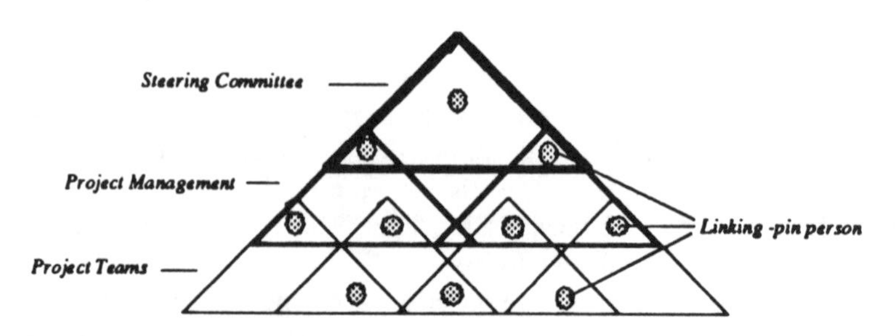

Fig. 10.3
Project Team Structure

IBM articulated three primary development strategies. These are described in more detail in Chapter 3.

- In the Monolithic or Waterfall strategy, all tasks associated with a system development phase (such as analysis) are completed before any of the tasks involved with subsequent phases are planned or undertaken.
- In the Version or Release strategy, after initial systems analysis or modeling, the product being developed is divided into subproducts and each subproduct is developed as a quasi-independent subproject. Two "sub" strategies are associated with the release strategy. The first is Sequential Release, in which one subproject is completed before the next subproject is started. The second is Concurrent Release, in which the subprojects are developed by separate teams often being scheduled in parallel.
- In the Evolutionary or Fast-Track strategy, the approach is to develop a full production version of the product as quickly as possible. The result is an undocumented, inefficient, and "dirty" production system that could be used while the Fast-Track development cycle is repeated to stabilize, add, or delete features based on the use of the production system. This strategy is the most commonly used in other industries for large or high-risk projects (for example, the new Parliament House and the Darling Harbour are using this strategy).

However, it is probably more practical to use a combination of all strategies for large projects. For example, the overall system development strategy for a

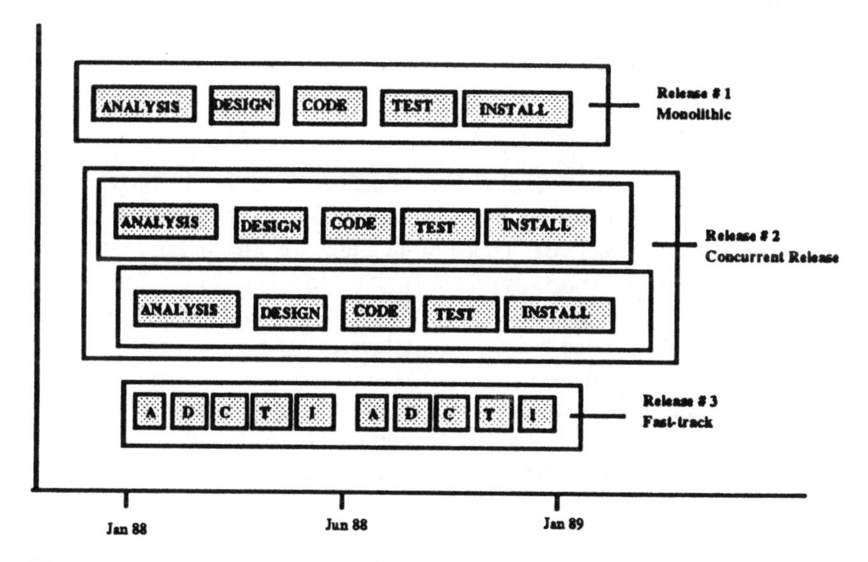

Fig. 10.4
Typical Large Project Development Strategy

large project is a Concurrent Release strategy with some of the high-risk subprojects being developed by means of Fast-Tracking, while more low-risk subprojects are broken into subreleases and developed using the monolithic strategy.

The use of multiple strategies within multiple releases adds further support to the requirement for vigorous and formal project management techniques. Apart from the sheer complexity of scheduling many people to many tasks, the project management system is vital in ensuring that changes in one subproject (release) do not have an impact on another subproject.

Quality Assurance

While formal quality assurance techniques (Technical Reviews, Walkthrus, and QA Groups) are apparently only partially implemented in the computing area of Australian companies, the need for Total Quality Assurance is essential in large projects.

As Jones' [op. cit.] research shows, the scale of effect issues have a significant impact in the area of defects in large projects. Simply, the level of defects per line of code is 200 percent higher for large projects than for small projects and up to 400 percent higher than for super-large projects.

This has serious implications for the intrinsic quality of the product, and the higher level of defect implicit in large projects adds an additional factor to the potential degradation of the requirements through analysis, data modeling, design, documentation, and coding defects. Jones [op. cit.] has found that the detection and removal of defects throughout *all* phases of the system development life cycle can account for 40 percent of the total development effort for large projects. In other words, without an effective Quality Assurance component, defects can account for around $2,000,000 of a typical large project's development cost!

Apart from the simple issue of defects, the relatively high incident of defects associated with large projects has a major impact on the project management issues. As discussed in Chapter 9, the "ripple effect" of defects passing between dependant project tasks is one of the biggest causes of poor estimation and project slippage. Of course, within a large project there will be hundreds of dependent tasks, and, as Boehm's classic model indicated, the cost of removing a defect rises roughly by an order of magnitude for each subsequent phase the defect remains undetected.

The solution to this issue is to enforce formal quality assurance techniques (Technical Reviews, Inspections, or Walkthrus) at the end of *each* task. As documented by Sprouster [1984] and Freedman and Weinberg [1982], it is cheaper and more effective to perform quality assurance in short, sharp sessions conducted by the development teams rather than the prevailing approach in computing (and many other industries, see Sprouster [op. cit.]) of large QA sessions conducted at the end of a number of tasks (subphase or phase reviews).

Finally, as already discussed, a feature of large projects is the use of staggered multiple releases of subproject deliverables into the production environment. While this is an essential feature of the Release and Fast-Track strategies, it presents a special problem in the area of production support and defect repair.

It is typical for the initial production releases of the project to be continually developed or enhanced as well as undergoing the normal post-production defect repair cycle. As a result, changes will be made to production data, data structures, function, and documentation. These changes may have a flow-on to the releases still in development. For example, a data structure may be shared between a production release and an in-development release. These changes must be reviewed not only by the production support teams but by the development teams working on related releases. In other words, production support QA must be shared by both development and production support teams. As a result, it is essential in large projects for the development and production support teams to be fully integrated and *colocated* to minimize communication breakdowns.

Support Tools and Technology

Substantial development has occurred over the past few years in the area of automated development support tools (particularly in personal computer-based tools). The development and support of large projects require dedicated access to the following tools:

1. CASE (computer-aided software engineering): These tools are currently in a period of rapid development. However, all of the widely implemented products (IEF, ADW, Excelerator) provide the ability to record data models, data flow diagrams, and design diagrams using graphic interfaces. CASE-type products reduce the effort required to document and maintain the various technical specifications across the various subproject teams.

2. Data dictionary: It is vital that the data being developed in the project is recorded in a centralized manner. Given that data is shared across the various subprojects (releases), any person on the team must be able to determine the logical and physical characteristics of the project's data and, in particular, which data is used in which subproject.

3. Desk-top publishing: Capers Jones [op. cit.] has data showing that documentation and paperwork consumes 30 percent of a large project's effort (compared with 10 percent for smaller projects). Just as CASE tools assist in reducing the effort in documenting the technical diagrams, the use of desk-top publishing tools assists in the production of user professional and operations manuals, forms production, and general system documentation.

4. Electronic mail system/groupware: As already discussed, a dedicated electronic mail system facilitates intraproject communication. Given the need for communication across many subteams and related specialist and management areas, this technology provides a quick, efficient, and auditable mechanism for recording and transmitting information. The use of cooperative and interactive software, such as Lotus Notes which enables many more people to share ideas, is useful.

5. *Project scheduling and tracking system*: As discussed later, change is a feature of large projects. The sheer size and complexity of project tasks, dependencies, and deliverables mandates access to tools for the fast microscheduling of tasks within each subproject and for tracking planned versus actual progress and costs. Total project planning could be achieved through the use of PC-based tools in a local area network or through the use of PC-mainframe technology.

6. *Electronic White-boards*: While this technology may appear ''low-tech,'' the standard electronic white-board provide 4–5 full screens and the ability to copy directly the content of each board. This technology is extremely useful in producing first-cut technical diagrams, project task lists, derivation of first-cut schedules, and critical path networks, as well as in the general recording of material in large team planning and technical sessions. The result can simply be copied and input into the relevant automated tools by a clerical support person.

7. *Switchable phone system*: The author's research indicates that the typical project person can lose up to 20 percent of a day in the activities of answering other project members' phones, taking messages, and locating that person. The use of a switchable phone system (or answering machines), by which each team member can switch his or her phone to a centralized message center staffed by a clerical support person, can lead to substantial gains in productive hours of work per day. In addition, the use of phone conferencing assists the electronic mail system in facilitating project team communication.

8. *Computer-aided instruction*: Because of the high organizational impact of large projects, there is a high demand for education for client groups associated with the project. Computer-based instructional packages (preferably mainframe based) can assist in the education of client user professionals by means of on-line tutorials, and the like.

Other support issues include application development and prototyping tools, automated testing and debugging tools, physical office design, rooms for Walkthrus and team meetings, access to technical libraries and, so on.

Stakeholder Management

As discussed throughout this handbook, a stakeholder is a group, organization, or person outside the direct organizational control of the project manager. In a small project, the stakeholders typically include the project sponsor, direct users of the project's deliverables, operations areas, and perhaps some internal or external software development and data administration consultants. In super-large projects, the stakeholder community exhibits the same diseconomy of scale as in the effort required for development.

In a super-large project, the stakeholders include groups both within the organization—such as Data Administration, Network, Operations, multiple client groups, and other dependent or interdependent projects—and, in most cases, areas outside the organization—such as vendors, other companies, unions, government and other major groups. For example, in one project with which the author was involved, over 60 external stakeholder groups were identified. This means that much of the project management effort (in excess of 50 percent) will be devoted to managing the stakeholder relationships in these types of projects. Further, the quality assurance processes, such as reviews, need to accommodate the requirement for multiple stakeholder reviews. For example, in a small project, a review of functional requirements may involve 5–7 people and take two hours. In a super-large project, a review of functional requirements may involve 20–30 people over a period of two days. Clearly, review techniques designed for small group review need to be modified for super-large projects.

The use of the linking-pin, as discussed, is another vital part of stakeholder management. In particular, linking-pins between the project and related projects ensure that the detailed technical issues (such as common function, data, design impacts, and so on) are managed at the appropriate level.

Team Building and Maintenance

Because of the long duration of super-large projects, team dynamics are considerably different than in smaller projects. In three super-large projects with which the author has been involved over the past three years, the team membership changed completely over the time for development. In one project, the only original team member left upon final delivery was a contractor!

Because of the inevitable turnover in super-large projects and other concerns (such as the need for high disciplined project management, multiple external relations, and large team size), team building and maintenance make up one of the key issues in super-large projects. As discussed by Constantine (1989) and Thomsett (1990), most computing organizations have failed to develop and implement effective models of teams; the processes of team formation and building are left to random selection based on technical skills and resource availability only. In the case of small projects, a randomly formed team may manage to produce the required system without having to face the morale and motivation crises that are a documented part of large projects. However, the team dynamics of super-large projects are such that there is a need to ensure that the requisite team roles as described by Thomsett and Constantine are present, and that the appropriate team maintenance and support mechanisms are built into the project.

Super-large projects require the deliberate and structured formation of a *project culture* that provides a common vision for the team members. The project vision is a vehicle for ensuring that the project's objectives and values are linked into the corporate objectives while providing a microclimate or culture that will sustain the project as team members leave and join throughout its life. In the super-large projects with which the author has been consulting over the past three years, this project culture is manifest in the project management documentation (scope, objectives, quality, and so on), as well as in "softer" concepts such as a project logo, project marketing material (brochures, portfolios), formal project induction programs, and regular (usually every three months) "timeouts," which combine formal planning and team-building exercises.

Development and maintenance of the project culture are key project management roles in super-large projects.

Change-Control Procedures

Probably the single most important feature of large projects is the previously mentioned scale-of-effect (see Figure 10.1). There is a nonlinear growth and effort pattern for large projects.

According to the Lines-of-Code in Figure 10.5, a change in initial requirements resulting in the need to provide an additional 5,000 LOC on a system of base size of 25,000 LOC requires an additional 2000 work hours of effort. To add the same requirement of 5,000 LOC to a large project of 125,000 LOC requires an additional 6000 work hours plus. This scale-of-effect impact is shown in Figure 10.5.

This leads to the formulation of Rob's Rule of Large Systems:

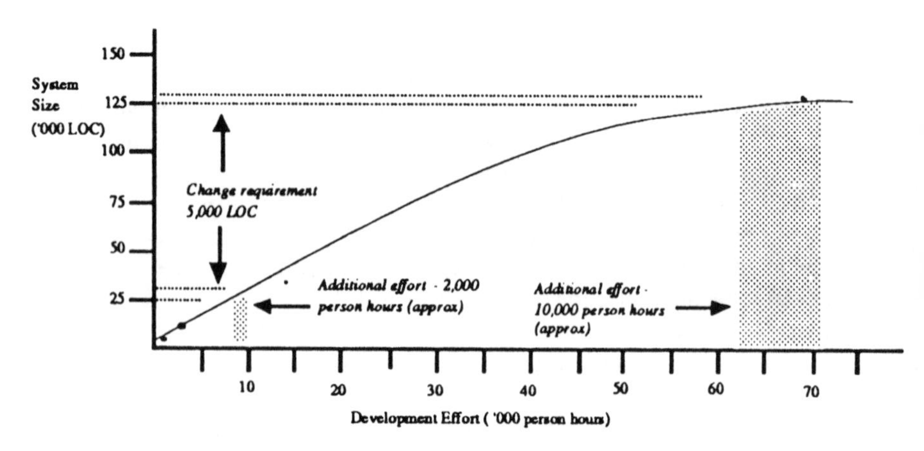

Fig. 10.5
Impact of Changes

For large systems, there is no such thing as a small change.

Capers Jones [op. cit.] refers to another IBM study, which indicates that the typical behavior of all information system projects is an expansion of requirements and effort by 50 percent over the project's development life. Given the scale-of-effect problem, the control of changes in large projects becomes mandatory.

In effect, any change to the project (requirements, costs, effort, staffing, and so on) should be treated as a mandatory "freeze" point and treated following the Change Control guidelines in Chapter 4.

Project Agreements or Contracts

As discussed, all large projects necessitate the use of multiple teams that cross and recross the existing inter- and intraorganizational boundaries.

It is typical for a large project to involve 20 or more separate teams—user professionals, computer professionals, specialist groups, unions, contractors, multiple vendors, and so on. In many cases, these teams are not in the direct scope of control of the overall project manager. Often key tasks on the project's critical path are under the control of an outside "subcontracted" area.

As a result, the development and commitment to deliverables, tasks, and deadlines involving teams outside the direct control of the project manager must be negotiated and formalized *prior* to the overall commitment by the project manager and the team to that specific part of the project. This is along the lines of the Project Agreement discussed in Chapter 5.

The formalizing of what is usually an informal review process is critical in large projects because it ensures that the project manager has considered the broader organizational support required and that the other areas are involved in developing a realistic schedule that recognizes their other ongoing work. Of course, all project contracts are reviewed by the Steering Committee prior to approving the particular component of the project development cycle.

Further, it is common in large projects that, given the large number of external groups or stakeholders involved in the project, over 50 percent of the project manager's time is devoted to managing these stakeholders. This involves negotiation, reviews, briefing, and general project management communication and "boundary riding." It also requires the assistance of Steering Committees as discussed in Chapter 6.

Summary

The management and development of large projects are critical issues in computing over the next decade. The relative lack of published material dealing with this area is symptomatic of the critcality of the issue.

To effectively develop and implement large information systems requires a reevaluation of traditional management techniques and attitudes. Senior management must become closer to the issues of project management. In their roles as project sponsors and Steering Committees they must adopt a "hands-on" approach to the project management and associated techniques and technology. In particular, the management of large projects requires special attention to the following areas:

- Project success.
- Project management technique.
- Project team organization.
- System development strategies.
- Quality assurance.
- Support tools and technology.
- Stakeholder management.
- Team building and maintenance.
- Change-control procedures.
- Project agreements or contracts.

Organizations can expect to gain major benefits from the new technology and delivered services and products involved with the next wave of information system development. However, if the implementation of the approaches discussed in this handbook be inadequate, many organizations will be facing not a "brave new world," but rather a "new nightmare" of system and organization failure of an unprecedented magnitude.

The hubris associated with large systems is well documented in the texts referred to in the Selected Readings at the end of this chapter. The failure of large projects is not inevitable, but the margin of success is narrow, and all levels of management must make the effort in time, education, finance, and leadership to ensure success.

References

L. L. Constantine, "Teamwork Paradigms and the Structured Open Team," *Software Development Conference*. San Francisco, 1989.

B. Dreger, *Function Point Analysis*. Englewood Cliffs, N.J.: Prentice Hall, 1989.

B. Boehm, Software Engineering Economics. Englewood Cliffs, N.J.: Prentice-Hall, 1981.

P. F. Drucker, *Innovation and Entrepreneurship*. New York: Harper & Row, 1985.

D. P. Freedman & G.M. Weinberg, *Handbook of Walkthroughs, Inspections and Technical Reviews,* 3rd ed. New York: Dorset House, 1982.

C. Jones, *Programming Productivity: Issues for the Eighties*. Los Angeles: IEEE Computer Society, EHO186-7, 1981.

P. G.W. Keen, "Information Systems and Organizational Change," *Communications of the A.C.M.*, Vol. 24, No. 1, 1981.

R. Likert, "The Principle of Supportive Relationships," *Organization Theory,* Pugh, D.S. ed. Middlesex: Penguin Books, 1971.

D. A. Ludlum, "$80M MIS Disaster," *Computerworld,* February 1, 1988.

R. Nolan, "Managing the Crises in Data Processing," *Harvard Business Review,* Vol. 57, No. 2 (March-April 1979).

L. Putman, "A General Empirical Solution to the Macro Software Sizing and Estimation Problem," *IEEE Transaction of Software Engineering,* Vol. SE-4, 1978.

J. Rothfeder, "It's Late, Costly, Incompetent—But Try Firing a Computer System," *Business Week* (November 7, 1988).

J. Sprouster, *Total Quality Control—The Australian Experience*. Sydney: Horwitz Grahame Books, 1984.

R. Thomsett, "Pragmatic Project Management—The Critical Issues for Senior Management," *Proceedings of ACC86, 1986.*

R. Thomsett, "Effective Teams: A Dilemma, A Model, a Solution," *American Programmer,* Vol. 3, Nos. 7–8, (July-August 1990).

Super-Large Projects: Selected Readings

Two excellent books on the behavior of large projects, systems, and high-risk technology are Professor Charles Perrow's *Normal Accidents, Living with High-Risk Technologies* (New York: Basic Books, 1984) and Professor John Gall's *Systemantics* (New York Times Book Co., 1977). Perrow's book deals with the failure of large projects in construction, engineering, nuclear power, and other areas. Gall's is a look at the behavior of large systems with a serious message but presented as a series of simple laws.

Three other books look at specific industries. Mary Kaldor's *The Baroque Arsenal* London: Abacus, 1983) details the massive problems in the projects that are developing new weapons technology, and Patrick Tyler's *Running Critical—The Silent War, Rickover and General Dynamics* (New York: Harper & Row, 1986) is an essential text as it presents both the political, human and project costs associated with the development of Trident submarines. Malcolm McConnell's *Challenger—A Major Malfunction* (London: Simon & Schuster, 1987) is similar to Tyler's book in that it presents a detailed account of the poor project management, political and human concerns, and massive communication breakdown typical of large projects under pressure of deadlines.

Management Disasters and How to Prevent Them by O. P. Kharbanda and E. A. Stallworthy (Brookfield, Vermont: Gower, 1986) provides a detailed examination of a large number of large projects including the Bhopal disaster and develops an excellent set of general diagnostics and preventative measures for large projects. Henry Petroski's *To Engineer Is Human* (London: MacMillan, 1985) is another broad-ranging discussion of the engineering problems associated with large projects and the role of management in preventing potential disaster.

While at the Australian Graduate School of Management (University of New South Wales), John Stringer produced a number of papers on the problems of managing large engineering projects. In particular, ''Management Problems of Large Engineering Construction Projects'' (Working Paper No. 82-005, May 1982) provides an extremely well researched approach to large project management, which has direct implications for large information system projects. The Auditor-General's Efficiency Report on the new Parliament House (Canberra: A.G.P.S., 1988) also contains many examples of the behavior of large projects.

Two other notable articles deal with large projects. One is ''Divad'' by Greg Easterbrook (*The Atlantic,* [October 1982], pp. 29–39), which records the massive problems associated with the Divad or Sergeant York automated all-weather gun platform. Jim Mintz's ''How the Engineers are Sinking Nuclear Power'' (*Science* 83 [June], pp. 78–82) contains an account of the 10:1 project cost and schedule blowouts associated with the second wave nuclear power plants in the United States.

Capers Jones' *Programming Productivity: Issues for the Eighties* (see References) provides articles written by some of the leading experts in software engineering. A number of these articles refers specifically to the projects in computing. Notable articles include ''On Understanding Laws, Evolution, and Conservation in the Large-Program Life Cycle'' by Lehman and ''Managment Perspectives on Programs, Programming and Productivity'' by Kendall and Lamb. Finally, Tracy Kidder's Pulitzer Prize-winning *The Soul of a New Machine* (New York: Avon Books 1981) details the development of Data General's Eclipse supermini and contains many examples of the behavior of large projects written with a journalist's flair.

11 Managing Object Orientation

Question: When is a paradigm not a paradigm?

Answer: When it is being marketed as a paradigm

For those computing people who lived through the Structured Revolution and the data-versus-function religious wars of the 1980s, there is a disturbing feeling of deja vu surrounding the object-oriented paradigm.

As Gerry Weinberg once commented in a 1981 seminar, "Everyone wants to see how the West was won but no one is interested in how it was made livable in afterwards." The structured revolution was driven by the development and technical concerns of improved analysis, design, and client involvement. As a person who taught and consulted in the 1970s and 1980s in structured techniques and project management, the author can fairly say that many "structured" organizations are still—in the 1990s—grappling with the managerial impact of those once new techniques. Many books were and are still being written on how to use structured techniques, on when to use structured techniques, and on whether rounded rectangles are superior to bubbles. Only a relative few, such as Ed Yourdon's "Managing the Systems Life Cycle" [1988] and Brian Dickinson's "Developing Structured Systems" [1981], were written on the management implications of the structured paradigm.

The sequence of technique first, management second, and organization last appears to be in danger of being repeated with the object-oriented paradigm.

While there is confusion about the meaning of the term "object-oriented," there appears to be an emerging consensus that the object-oriented approach involves a number of key elements:

- *Abstraction*: The defining of data by deriving common attributes across an entire group of data.
- *Encapsulation*: The absorption of a data structure and function within a single (albeit) complex module.
- *Inheritance*: The "copying" of features contained in one object into related objects.
- *Reusability*: The ability to maximize the reuse of objects with minor modification.
- *Integration*: The capability of sharing object/data with minimum physical redundancy.

While this rather simplistic view of the object-oriented paradigm will probably upset some object-oriented gurus, it provides a base for exploring the significant issues implied in ODD (object-oriented development) for the management of information systems. Brian Henderson-Sellers [1991] provides an excellent "model"-independent overview of the object-oriented approach.

In particular, OOD will need reevaluation of and, in some areas, completely new models for:

- Project management.

- Team roles.
- Quality.

Project Management Impact

Contemporary project management models covered in this handbook incorporate issues such as methodology, metrics, and project context. Most project management environments have now been modified to include the impact of structured techniques. For example, packaged methodologies such as Method/I and SDM have evolved from unstructured to structured methodologies. Similar modifications will be required for OOD.

Estimates Impact

It is now widely accepted by proponents of OOD that to optimize the concepts of abstraction, encapsulation, and inheritance will require even more emphasis on the analysis and design phases of the system development life cycle. Douglas Dedo of Hewlett-Packard [1991] suggests that the analysis and design phases in OOD will require 100 percent additional effort as compared with traditional development approaches. In larger object-oriented projects, there is also an additional effort in system integration and testing since the common objects may have been modified and need to be reintegrated. As shown in Figure 11.1, the detailed design, programming, and testing of modules should require less effort than other development methodologies.

System Development Strategies

The inherent linear sequencing of the logical activity of data/function analysis and the physical activity of design associated with structured and information engineering development techniques will become highly interactive in OOD, and there will be blurring of the analysis and design boundary. This blurring is being additionally supported by the highly interactive nature of CASE, JAD, or "rapid prototying" development technology. As a result, the OOD methodologies, such as Meilir Page-Jones and Steven Weiss' Synthesis [1989], will even more "front-loaded" in terms of effort, complexity, and risk. The use of the more dynamic system development strategies (see Chapter 3) such as Rapid Application development, Concurrent Releases, and Fast-Tracking are required to accommodate the iterative development nature of OOD.

Metrics

An increasing number of organizations have begun to develop metric models based primarily on IBM's Function Point and source lines of code. As both these metric models are based on traditional life cycles or methodologies, they

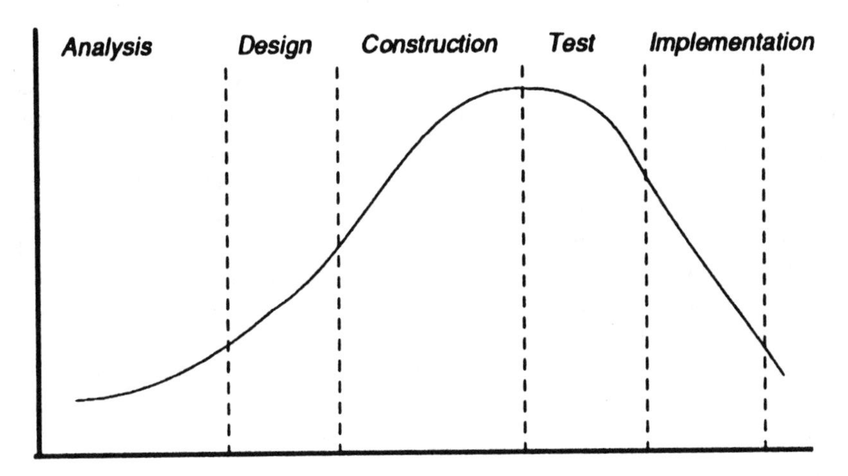

Traditional life-cycle staffing model

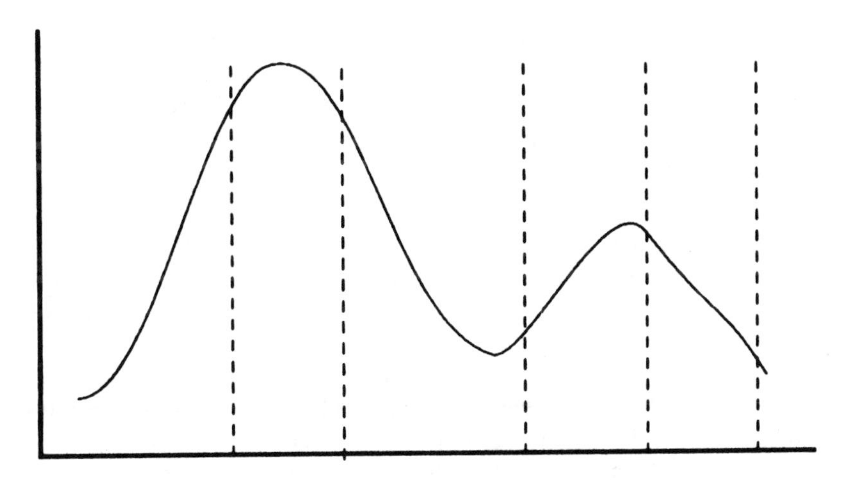

Object-oriented life-cycle staffing model - speculative

Fig. 11.1
Project Staffing Profiles

will provide little assistance in either estimation of OOD or productivity metrics. For example, Function Point is based on the premise that a single logical input results in a single function, not an instance of a complex object. As reported in *American Programmer* [Summer, 1989] the early comparative measures of OOD versus structure development indicate radical differences in total effort and dispersion of effort across the life cycle. New metric models must be developed for OOD.

Project Boundaries

A key element of contemporary project management is the recognition that projects exist within a complex network of clients, project sponsors, internal support groups such as data base administration and network support, interdependent projects, and other stakeholders. Inherent in the OOD paradigm are the concepts of sharing, commonality, reusability, and corporate versus project ownership of objects. The management of these complex project boundaries and interfaces (both organizational and object) will demand a higher degree of sophistication and formality in key project management concerns such as planning, charge control, and risk management.

In larger OOD projects, at least three distinct teams can be identified in addition to the number of subrelease teams associated with large projects (see Chapter 10):

- *Application development*: These teams develop the business or presentation layers of the software that is, the user layer.
- *Object development*: These teams develop the objects that support the development teams and are focused in the general application of the objects as distinct from the specific business application being developed.
- *Object repository*: These teams manage the physical shortage of the objects in the organization's object repository.

These interfaces and new boundaries are shown in Figure 11.2.

The need for these additional specialist teams increases the complexity of the already complex process of stakeholder management described in this handbook.

In a study of a series of major object-oriented projects, Thomsett Associates has found that it was the stakeholder and boundary issues rather than technical problems that lead to severe project problems. In particular, the focus of the object development and repository teams is often different from that of the application development teams. The support teams are concerned with broad organizational issues such as reuse and generalization, while the application teams are concerned with timely deliverable of project requirements. Without tight project management and stakeholder negotiation, many of the benefits of OOD will be lost to the imperative of tight deadlines and cost containment.

Team Roles Impact

Associated with the complexity of management of organization and object boundaries will be the emergence of two new project team roles. As already argued by

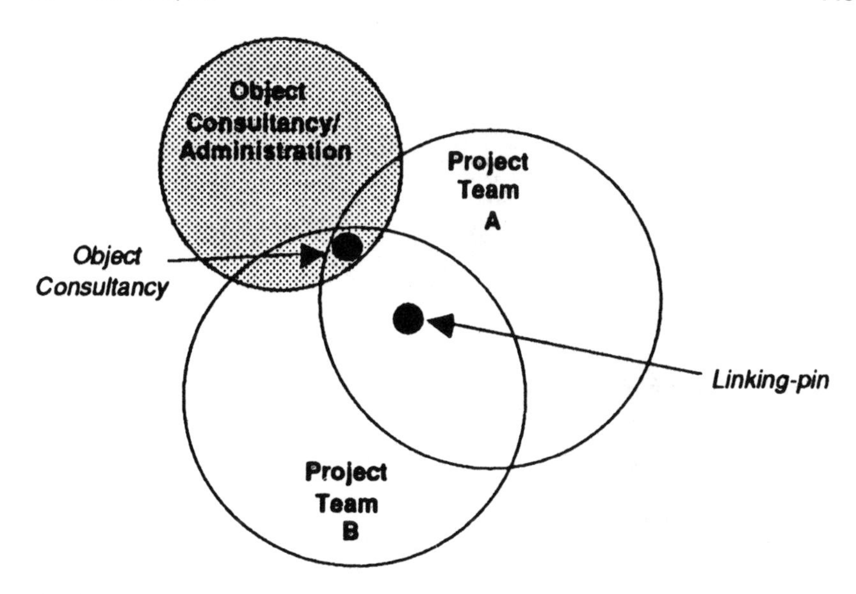

Fig. 11.2
New Boundaries

Constantine and Thomsett [op. cit.], organizations have tended to ignore the need to systematically examine the key roles required for an effective project team. Contemporary project team models recognize that the hierarchical (or closed) project team model does not provide the flexibility (both internally and externally) required to deal with the complex interproject boundaries required for structured system development. As detailed earlier, OOD will require even higher levels of interteam communication and cooperation.

To deal with this, OOD will require specific focus on two new team roles:

- Linking-pin.
- Object consultancy.

Linking-Pin

The concept of the linking-pin role was first articulated by Renis Likert [op.cit.]. Likert proposed that, in contrast to the primarily up-down communication patterns of traditional hierarchical organization structures, the sharing of team members across team boundaries would provide a valuable horizontal communication path.

The need for a person with a team whose primary focus was the communication across teams at a detailed technical level was applied by Thomsett [1988] and has been implemented in a series of major current OOD projects.

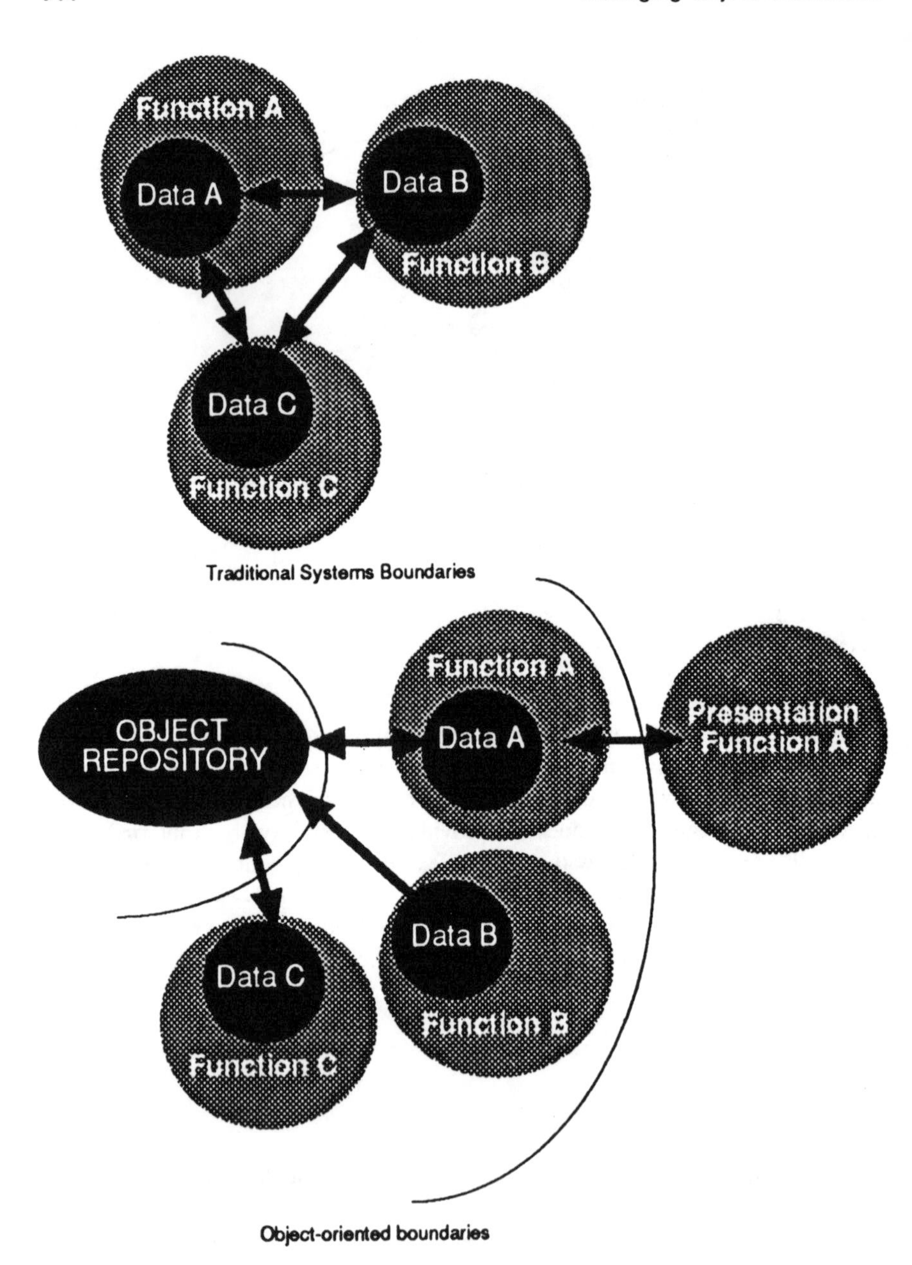

Fig. 11.3
New Team Roles

The linking-pin role is shared across a number of teams. For example, two teams involved in developing systems using common objects (data and/or function) would jointly sponsor a person who would live in the two teams. This linking-pin person would provide a more detailed interface in support of the more overall interface issues traditionally undertaken by the project manager.

Object Consultancy

While many structured organizations have created mature and centralized Data Administration and Database design groups, the very nature of OOD demands a higher level of consultancy in the area of object management. In structured development, the logical modeling of data and function is undertaken by analysts in the team, and the physical modeling of data is undertaken by the centralized Data Administration and Design groups. Because of the highly interactive nature of the logical/physical boundary in OOD, centralized support groups are generally not able to provide "adequate" consultancy. Even with a fully integrated object library or repository to assist in object management, the level of complexity in evaluating objects, establishing the variances between inherited objects, etc., OOD teams need resident "object consultants" drawn not only from the DA and DB groups but from the application areas. These object consultants provide internal consultancy on object management, as well as linking to the DA and DB areas. They supplement the traditional system analysts and designers in the team.

Quality

One of the significant by-products of the increased competitive, financial and technological pressures of the 1980s is the emergence of high-risk projects with fixed deadlines. The "fixed deadline" syndrome has often lead to compromises in project scope and quality. The degradation of project quality to meet fixed deadlines, constrained people and financial resources, and changing client requirements has become accepted reluctantly by most computer teams.

As described in Chapter 9, software quality is generally accepted to include the following attributes:

- *Conformity*: The meeting of agreed-upon data and function requirements.
- *Reliability*: The system performs with consistency and correctness.
- *Maintainability*: The system is easy to maintain.
- *Reusability*: The system maximizes the reuse of common function and data.
- *Efficiency*: The system performs with optimum response, resource usage, etc.

- *Portability*: The system is capable of operating in different environments.
- *Usability*: The system is easy to use from the client's perspective.
- *Auditability*: The system is auditable.
- *Security*: The system is secure from unauthorized access.
- *Flexibility*: The system can be enhanced easily.

In a survey conducted by Thomsett Associates with over 600 analysts, designers, and programmers, the impact of the past decade was disturbingly evident. While more than 95 percent agreed that conformity, reliability, maintainability, and flexibility were essential attributes, only 10 percent considered reusability and portability as essential attributes. In addition, concurrent and extensive surveys conducted in one major Australian organization revealed that business groups perceived that conformity, reliability, and usability were their concerns. Maintainability, reusability, and portability are transparent to most clients! Further, both groups accepted that deadlines were a de facto measure of quality.

In other words, the primary OOD concerns of reusability and portability are at risk in the current economic environment. As described by the proponents, OOD must face the current reality of contemporary pragmatic definition of quality. As long as there is a continuation of the concentration, in the broader organizational sectors, on short-term returns as documented by Garvin and Hayes [1982], project managers in object-oriented projects will be constantly faced with a compromise between the long-term gains of object orientation and the short-term goals of projects.

Perhaps the single biggest issue facing the emergent OOD paradigm is its very nature. OOD challenges long established organization and application boundaries. The structured techniques historically recognized the system boundary and interface issues via such diagrams as the Context or Level Zero data flow and the subject data model, and basically worked with traditional application boundaries. The effective implementation of OOD not only requires a new development paradigm but also an associated shift in organization political paradigm since the often parochial data ownership by business groups (and their systems) will be challenged by OOD (see Figure 11.2).

Unless these serious management issues are addressed, the OOD paradigm will likely remain the concern of software engineers and theorticians locked out of the main stream by overriding organizational concerns of short-term deliverables and territorial politics.

Because of its very power, the OOD paradigm is an organization paradigm first and a system development paradigm second.

References

L. L. Constantine & R. Thomsett, ''Building Effective Project Teams,'' *Workshop Notes*. Canberra: R.T.&A., 1990.

D. Dedo, ''Object-oriented Data Bases,'' *American Programmer*. New York: American Programmer, Inc., Vol. 4, No. 10, October 1991.

B. Dickinson, *Developing Structured Systems*. Englewood Cliffs, N.J.: Prentice-Hall, 1981.

D. Garvin & R. Hayes, ''Managing As If Tomorrow Mattered,'' *Harvard Business Review* (May-June 1982).

B. Henderson-Sellers, *A Book of Object-oriented Knowledge*. Englewood Cliffs, N.J.: Prentice Hall, 1991.

M. Page-Jones & S. Weiss, ''Synthesis: an Object-oriented Analysis and Design Method,'' *American Programmer*, New York: Children's Computer Company, Vol. 2, Nos. 7–8, Summer 1989.

R. Thomsett, ''Managing Super-large Projects: A Contingency Approach,'' *IBM Share/Guide*. Sydney: February 1988.

E. N. Yourdon, *Managing the System Life Cycle*. Englewood Cliffs, N.J.: Prentice Hall, 1988.

Function Point Tutorial

Counting is wonderful
Count Dracula
(Sesame Street)

Introduction to Function Point Estimating

Function Point is a software sizing, productivity measurement, and estimating technique developed by Alan Albrecht [1979] and others in the late 1970s. It is based on two assumptions:

- The complexity and size of a software system are major determinants of the length of the development process.
- The complexity and size of a software system can be derived by examining and counting the data complexity and volume.

As such, Function Point counts have been shown to be a reasonable method for deriving ballpark estimates independent of more problematic software measures such as lines of code. Capers Jones [op. cit.] identifies at least 10 variations on counting lines of code (LOC). Further, unlike LOC, Function Point counts can be estimated as early as the first phase of the system development process.

However, it should be emphasized that Function Point techniques ignore the differences between individual programmer or analyst productivity and quality requirements, and are not suitable for systems with complex algorithmic processing or hardware or telecommunications intensive projects.

As a result, Thomsett Associates recommends that Function Point counts and associated estimates are used to confirm the estimates developed by other techniques such as Wide-Band Delphi.

Function Point sizing is carried out by:

- Defining the data attributes, called Application Elements or Functions, of an application.
- Assigning a point score value to these functions.
- Applying a Complexity Weight to the functions.
- Identifying and applying Processing Complexity factors of the system.
- Calculating the total Adjusted Function Point score for the system.
- Deriving an estimate using Adjusted Function Pointcount and organization Productivity Index, which is an historical metric based on the hours/Function Point to develop or enhance systems.

Function Points are based on the user's external view or logical model of the application. Function Points are established for an application by listing, classifying, and counting different elements of an application's data. It should be noted that the use of the word "function" is somewhat misleading as the counting is based on data, not function. By counting data, Function Points de facto count processes, since processes exist to process data.

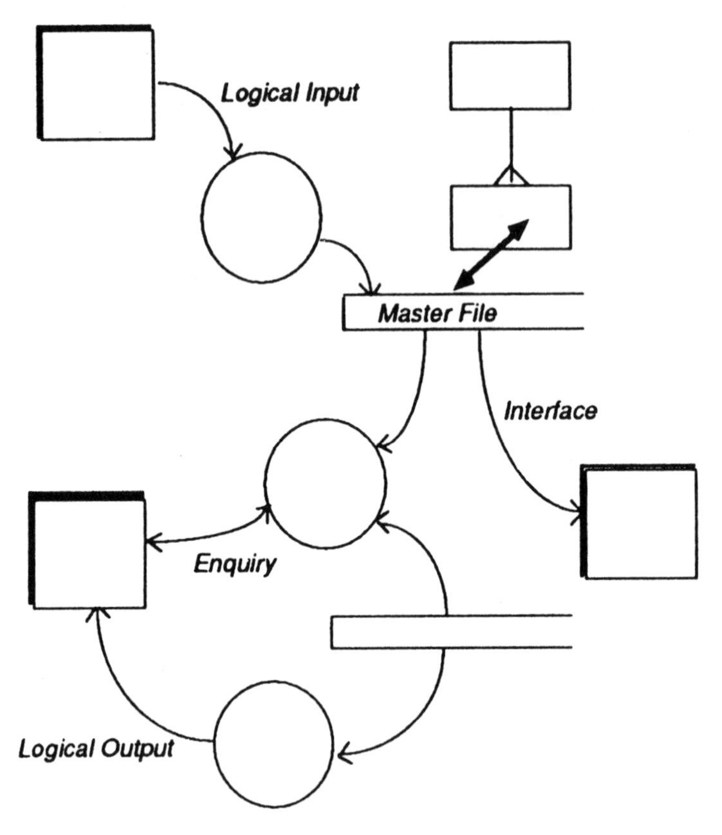

Fig. A.1
Data Flow Diagram/Entity-Relation Model

As shown in Figure A.1, data flow diagrams and data models are ideal vehicles for deriving Function Point counts as they depict graphically all data components.

This Appendix is designed to introduce Function Point at a level that can beused to calculate a "first-cut" Function Point count for developments and enhancements.

For a more detailed description of the technique, Brian Dreger's *Function Point Analysis* [1989] provides a complete and standard approach to this technique.

Developing a Function Point Estimate

Function Point estimation is based on counting five Application Elements or Functions for the new system or enhancement. It is important to understand that the counting is based on logical, not implementation-based or physical, functions. For example, the same data may be input to the system via an on-line screen and a batch update. Provided the processing of both inputs is the same, the data would be counted as one logical input although it exists in two different physical formats.

Functions

Business functions consist of the following:

- *External inputs* to an system, such as input screens or forms.
- *External outputs* from an application, such as output screens or reports.
- *Logical master files* master files/user views used in the system.
- *Enquiry transactions* inputs that result in a immediate output.
- *Interfaces* data to/from other applications that form application dependencies.

Calculating Function Points

Points are scored for each function identified in the proposed system, and these scores are then adjusted for the intrinsic complexity of each function. It is important to avoid duplicated counting. Each business function is counted once and once only. Details of the various functions and the complexity factors, along with guidelines on their use, are found in later sections of this appendix.

After analysis of the system components and functions, a count is made of the number of each type of the preceding elements under a subclassification of complexity, that is, simple, average, or complex. These elements are then weighted by a relative complexity weight for each of the three subclassifications. These weights have been established from experience by IBM and many other organizations over a large number of projects and have been accepted unaltered by other users of the Function Point technique. The *Unadjusted Function Point* score is then calculated by multiplying each function by its complexity and adding the results together.

Further, analysis is then made of the processing complexity of the system, a score value estimated for each and these factors applied as a further weighting to the point score to arrive at a total *Adjusted Function Point* sizing.

Processing Complexity (currently 14 specific characteristics have been identified) are factors in a project that, if present, can influence the overall complexity of the project. An example is the extensive use of communications facilities in the system. These processing complexity factors are a subset of the factors in System Complexity in Thomsett & Associates' Risk Assessment model (Chapter 8).

Function Point Index or Function Point Productivity

The index is a value expressed in hours-per-function point. The Productivity Index, derived historically for the organization, is then applied to the Adjusted Function Point score to arrive at an estimate for a particular project.

This estimate is an *actual* hours (WE) estimate for the proposed system. This, of course, must be followed seperately by an *elapsed* time (ED) calcuation based on resourses available and other external factors (see Chapter 3).

Variety of Function Point Indexes

In analyzing systems and calculating function points, to arrive at an organization's Productivity Index, several indexes should be calculated. They could cover a variety of technology platforms and, in particular, language-class used in the organization.

Experience from people such as Capers Jones [op. cit.] and Dreger [op. cit.] indicates substantial variations in index values according to the programming "language" or language class used. Assembler languages indicate an index of 20–30 hours per function point; COBOL systems indicate an index of about 10–20 hours per function point, while fourth generation languages and application generators such as SQL, TELON, Natural, etc. appear to have an index value only half of that for COBOL. Caper Jones has developed a detailed model with in initial variances among the five classes of development languages.

These values may be used as a basis for estimating future workloads. However, research on function points has shown that there is considerable variance in function point productivity across organizations. For example, one organization with which The Thomsett Group works uses 40 hours per function point in COBOL as against the industry average of 10–20 hours per Function Point. This variance is due to the difficulty of measuring many of the factors that affect productivity and, in particular, the massive variations in individual productivity between people. Remember, Risk Assessment contains some 60-plus factors, while Function Point technique includes only 14 factors as Processing Complexity factors.

Ideally, each organization should develop its own metric data base reflecting the history of their own projects and productivity rates. This is done by counting the function points delivered by effort for a number of projects, recording the Function Point size, Processing Complexity, quality requirements, risk, language class, and any team member skills that can be identified as unique to the project. These project histories can then be used to derive an organization-specific set of Productivity Indexes.

Following the project planning and documentation processes covered in this handbook would automatically provide sufficient metrics to facilitate the collection ot this data without additional effort from the project teams.

Measuring the Work Product Output

Beause the end result of Function Point estimating is a total of person-hours of work, it may be regarded as a measurement of effort and cost.

It is valid to associate Function Points with the value of an application to the user in the sense that they represent the amount of function delivered, that is, user facilities. There is no relationship to any regard of return on investment expected from the application (see Chapter 7).

Enhancement to Existing Applications

A variation of the Function Point technique can be used for enhancement projects as distinct from new developments. Given that enhancements involve alteration to an existing production system, the technique involves developing an Adjusted Function Point score for the production system and then counting the number of new, deleted, and altered functions involved in the enhancement. Dreger's book [op. cit.] provides a detailed discussion of this variation.

Definitions and Guidelines

The following section provides basic definitions of the functions involved in function point sizing and the complexity weights for each function type. These will enable the production of an initial sizing of the system, which can then be used to develop an initial estimate using the organization's Productivity Index.

User External Inputs

User external inputs are counted for each one-way user input to the system across the system boundary that requires a business function from the users view:

- Count *all* the screens used to input each transaction. These transactions cover user transactions and should not include any software system generated screens or control screens;
- Count all other unique inputs, such as OCR, MICR, and hard copy documents. An input is considered to be unique if it has a different format or requires different processing logic from other inputs. Each transaction type should be listed separately.
- Change, add, and delete transactions are counted separately if significantly different processing logic is required for each.
- The same screen used for input and output should be counted under both inputs and outputs if different processing logic is required.
- Do not list inquiry screens under this category.

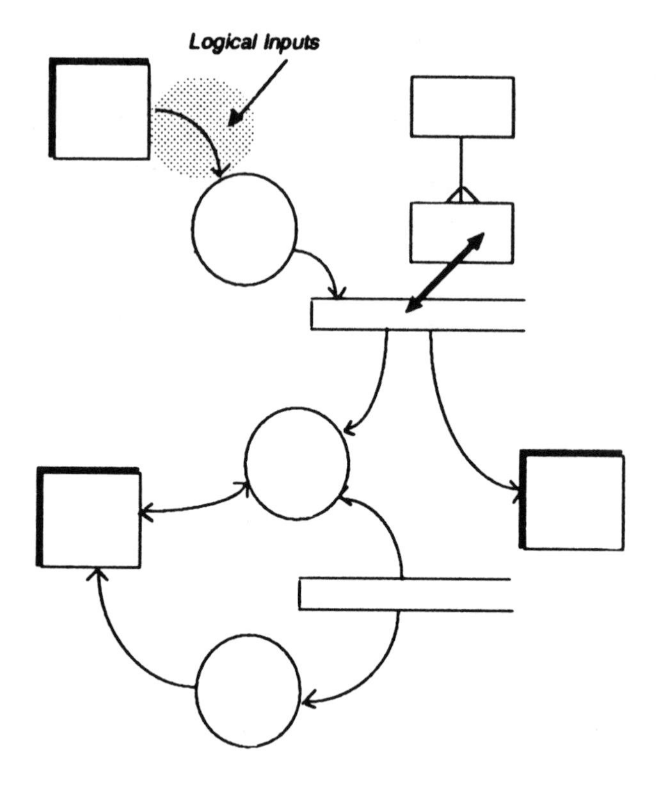

User External Outputs

User external outputs are data generated inside the system and dispatched one-way to the user across the system boundary:

- Count all unique external outputs. It is considered unique if it has a format that differs from the others, or if it requires unique processing logic to provide or calculate the output data. The same report for two different destinations is counted as only one, while a detail report and a summary report are counted as two.
- Count each output terminal screen that provides data, error, or operator messages, unless it only echoes what was input plus the error message, in which case it is not counted.
- Count all batch error reports, batch system reports, and other unique outputs, such as MICR and graphics if they are provided.

An inquiry is a simple search for specific data usually using a single search key. List each input/output combination where the on-line input generates an

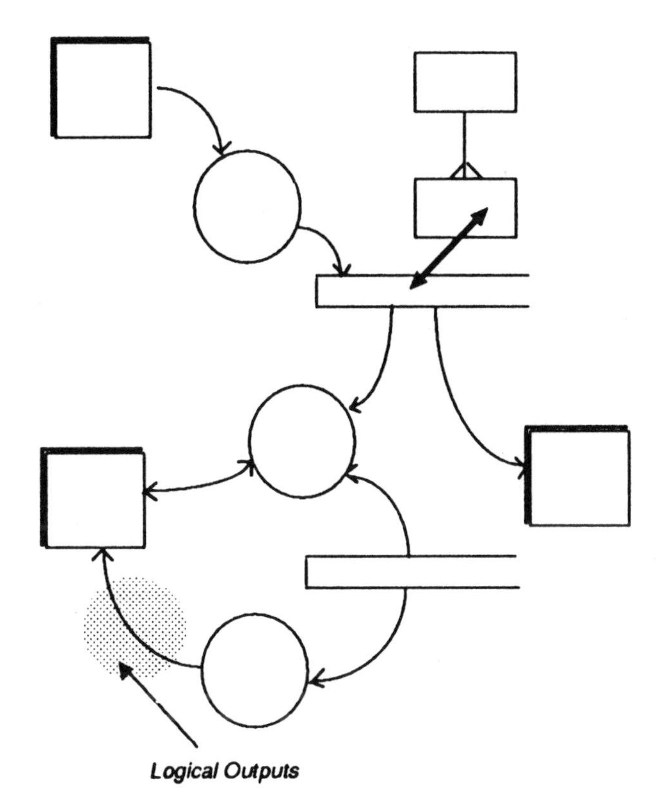

immediate on-line output. No *update* of the accessed file/s is involved in an inquiry:

- Count each unique formated or uniquely processed enquiry that results in a file search for specific information to be presented as a response to that enquiry.
- Count any major query facility that provides a hierarchical structure of inputs, outputs, and inquiries to handle many keys and many operations as the appropriate classification (that is, each input, output, and inquiry is counted separately).
- When data from an enquiry is subsequently used as an update, list it as both an enquiry and as an input.

Function Complexity Weights

Complexity ratings are used to distinguish the relative difference of processing or programming complexity in like system functions. Simple reports, for example, require less complex programming than a report with many fields and subtotals.

Complexity is divided into three basic categories:

- *Simple*: The minimum level.
- *Average*: The normal level.
- *Complex*: The high level.

Figures A.2 and A.3 provide the standard weights for the five function types.

For inquiries, the more complex of the input or output component is the one used to develop the Function Point score. For example, if the input component of the inquiry is simple (complexity weight of 3) and the output component is complex (complexity weight of 7), then the output component complexity weight (7) is used for the inquiry.

The scores are transferred to the Summary Sheet at the end of this Appendix (Figure A.4).

User Logical Master Files

List each unique machine-readable logical file, and, in the case of data bases, each logical grouping of data. These logical groupings are from the point of view of the user and not from the way the data base has been constructed. To assist in your decision, consider how many files would be needed in a manual implementation of the system.

Generally, the process of data analysis will provide logical file structures with user access keys; each unique key indicates a unique logical file:

- Count a file only once even though it may be used in several steps of a multistep job.
- Count logical files only. In the case of an indexed sequential file this is only one file as the index files related to the data file are maintained by the monitor system.
- Count any intermediate/temporal files or sort files.
- For complex integrated management system data bases, list one logical file for each user view used by the application.
- Count hierarchical paths viewed through logical relationships or secondary indices as additional logical master files.
- When logical master files are shared between applications, one must be designated as the owner and the other would list it as an interface file.
- Do not count batch files or temporal files used to generate outputs or to simply temporarily store inputs.

Input Complexity	1-4 attributes	5-15 attributes	16 + attributes
0 or 1 files accessed	3	3	4
2 files accessed	3	4	6
3 + files accessed	4	6	6

Complexity Weight : Simple - 3 Average - 4 Complex - 6

Output Complexity	1-5 attributes	6-19 attributes	20 + attributes
0 or 1 files accessed	4	4	5
2 or 3 files accessed	4	5	7
4 + files accessed	5	7	7

Complexity Weight : Simple - 4 Average - 5 Complex - 7

File Complexity	1-19 attributes	20-50 attributes	51 + attributes
1 logical record/ entity	7	7	10
2 - 5 logical records/entities	7	10	15
6 + logical records/entities	10	15	15

Complexity Weight : Simple - 7 Average - 10 Complex - 15

Fig. A.2
Complexity Weights (Input, Output, File) [Dreger 1989]

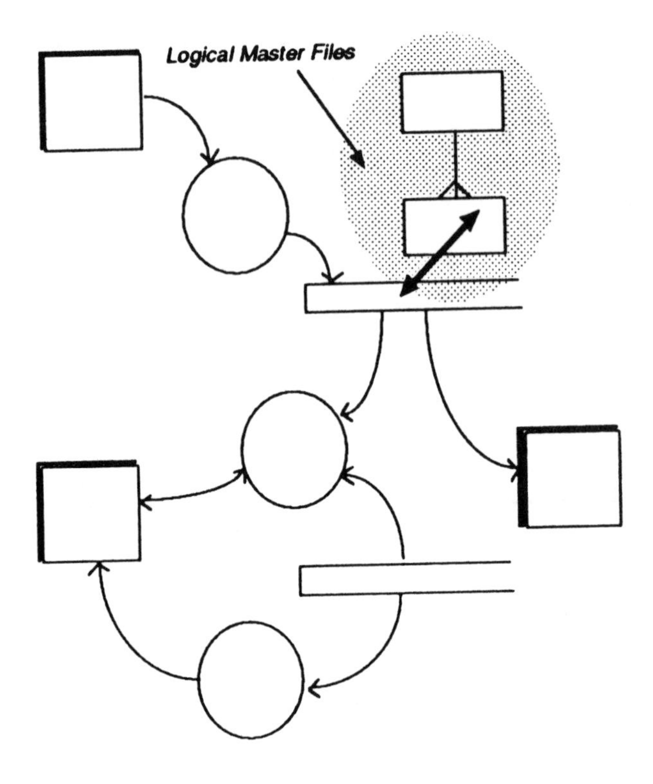

Interfaces to Other Systems

Interface files are files created solely to be passed to other applications as an interface to that application. Input files that have not already been counted by the initiating system are also classed as interfaces:

- Count logical master files used in this application but created and maintained in another application as an interface file.
- Count logical files of transactions or individual transactions used in this application and passed on to another application as an output from this applicatoin and an input for the other application.
- Count any files that are copied for distribution to one or more other systems as only one interface file.

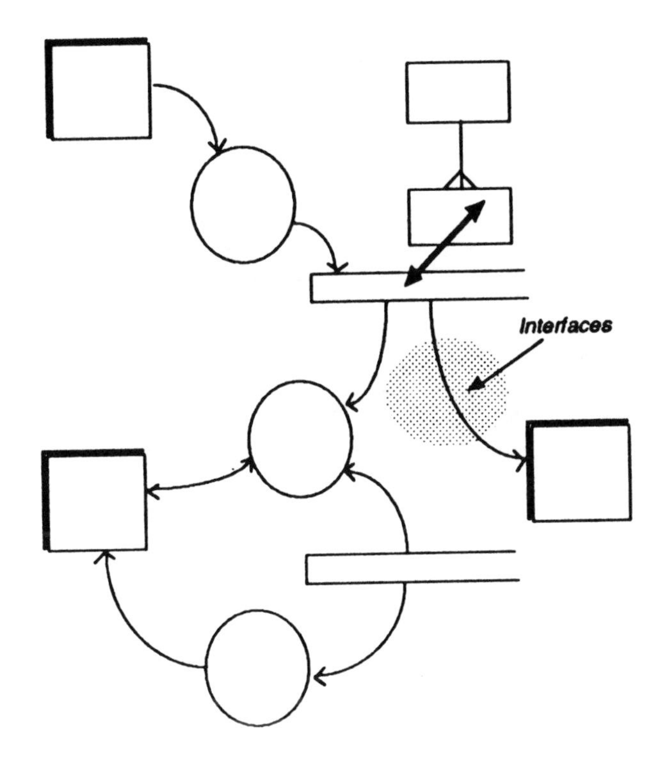

User External Enquiries

Processing Complexity

The Summary Sheets are used as a means of quantifying the influence of key characteristics or risk factors of a project. In Function Point sizing, some 14 characteristics have been identified as influencing the processing complexity of a project. These are termed Processing Complexity or Application Characteristics.

There are six degrees of influence (risk) in each processing complexity factor. After reviewing each characteristic, assign the value for the degree of influence that the characteristic has on the development, or the enhancement, of the application.

0 = Not present, or no influence.
1 = Minor (insignificant) influence.
2 = Moderate influence.

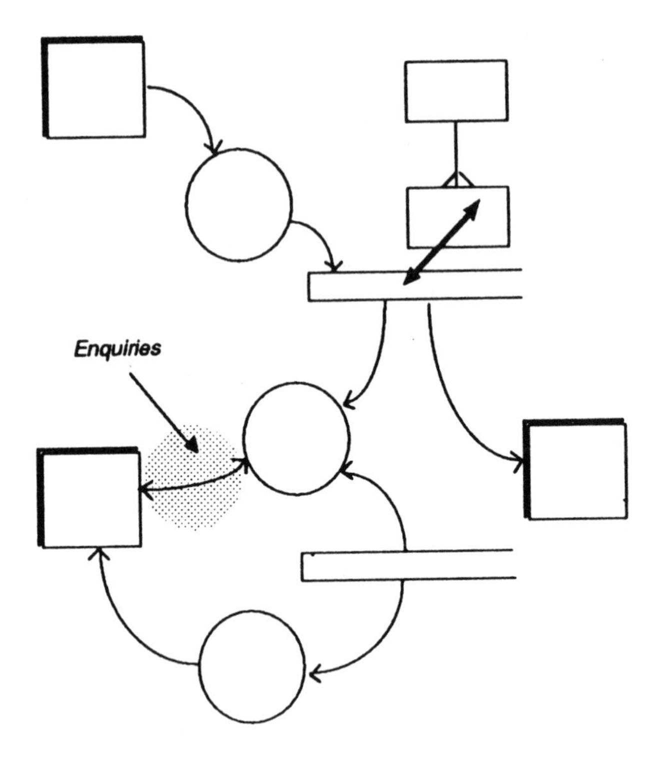

3 = Average influence.
4 = Significant influence.
5 = Strong influence throughout.

Dreger [op. cit.] provides a more quantified model of processing complexity but, for initial estimates in a project, a subjective assessment of the level of influence of each processing complexity factor on the project is usually more appropriate since the details of the technical issues have not usually been finalized.

Processing Complexity Factors

1. *Data communications:* Data used in a project is sent or received over communication line facilities (0 = batch, 5 = completely on-line data interfaces).
2. *Distributed function:* Distributed processing functions or distributed data are a part of the application (0 = totally centralized in one application, 5 = distribution throughout the network).

Enquiry Input Complexity	1-4 attributes	5-15 attributes	16 + attributes
0 or 1 files accessed	3	3	4
2 files accessed	3	4	6
3 + files accessed	4	6	6

Complexity Weight : Simple - 3 Average - 4 Complex - 6

Enquiry Output Complexity	1-5 attributes	6-19 attributes	20 + attributes
0 or 1 files accessed	4	4	5
2 or 3 files accessed	4	5	7
4 + files accessed	5	7	7

Complexity Weight : Simple - 4 Average - 5 Complex - 7

Interface File Complexity	1-19 attributes	20-50 attributes	51 + attributes
1 logical record/ entity	5	5	7
2 - 5 logical records/entities	5	7	10
6 + logical records/entities	7	10	10

Complexity Weight : Simple - 7 Average - 10 Complex - 15

Fig. A.3
Complexity Weights (Enquiry, Interface) [Dreger, 1989]

3. *Performance:* Performance in either response times or throughput is a consideration in the design, implementation, and maintenance of the project (0 = standard performance, 5 = high performance is essential).
4. *Heavily used configuration:* The user wants to run the application on existing or committed hardware that will be heavily utilized (0 = no operation restrictions, 5 = heavy constraints).
5. *Transaction rates:* The transaction rate is high and has influence on the design, implementation, and maintenance of the application (0 = standard response times, 5 = response times are critical).
6. *On-line data entry:* On-line data entry is provided in the application (0 = no interactive data entry, 5 = interactive data entry of >30 percent of data).
7. *End user efficiency:* The user requires simple, GUI interfaces and ease-of-use (0 = no specific requirement, 5 = full GUI interface).
8. *On-line update:* The data bases or master files are updated on-line (0 = no on-line update, 5 = high volume update with recovery considerations).
9. *Complex processing:* Internal functional processing is complex. It is also considered complex if there is a requirement for complex exception processing resulting in many incomplete transactions that must be resolved (0 = simple processing, 5 = complex logic, security, exception conditions, and input/output formats).
10. *Code reusability:* The application code is to be specifically designed, implemented, and maintained to be reuseable in other applications or at other sites. Use the allocations in the following table:

		%		%
0	=	0	to	10
1	=	10	to	20
2	=	20	to	30
3	=	30	to	40
4	=	40	to	50
5	=	50	to	100

11. *Conversion/installation ease:* Conversion and installation ease are to be incorporated into the design and implementation. A conversion and installation plan was provided and tested during system test (0 = no special considerations, 5 = user specific conditions and requires tools).
12. *Operational ease:* Operational ease is to be incorporated in the design and implementation. Well designed start-up, back-up, and recovery procedures were provided by the project and were tested during system test. The application minimizes manual activity in the machine room (0 = no special requirements, 5 = unattended operations).

13. *Multiple site installation:* The system is to be specifically designed, implemented, and tested, to be installed at multiple sites (0 = single site, 5 = multiple sites).
14. *Facilitate change:* The application has to be specifically designed, implemented, and maintained to facilitate change and ease of use by the user. Flexible inquiry capability is provided for the user. Business information and parameters subject to change are grouped in user maintainable tables and not embedded in the code (0 = no special requirements, 5 = control data and key tables maintained by users).

A Final Note on Funtion Points

Function Points have been adopted by most major computing organizations because they easily provide a standard and technology-independent method of sizing software products. However, Function Points are a software sizing technique and, as such, do not measure the nonsystem effort involved in developing most information systems (see "Scope" in Chapter 9). As a result, Function Points should be treated as a ball-park sizing technique and used to confirm that the development effort estimates (software and nonsoftware) created through work breakdown structures and Wide-Band Delphi are in an acceptable range.

Finally, Function Point estimation is a "total develoment cycle" model and does not provide phase estimates and task breakdowns. These have to be done for scheduling using the other estimation techniques described in this handbook.

References

A. Albrecht, "Measuring Application Development Productivity," *Proceedings Joint IBM SHARE/GUIDE*, October 1979.

B. Dreger, *Function Point Analysis*. Englewood Cliffs, N.J.: Prentice Hall, 1989.

T. C. Jones, "A Language Level to Fit the Job," *Computerworld* (December 9, 1988), pp. 21–24.

Business Function	Number	Complexity	Weight	Line Total	Type Total
Inputs		Simple	x 3		
		Average	x 4		
		Complex	x 6		
Input Total					
Outputs		Simple	x 4		
		Average	x 5		
		Complex	x 7		
Output Total					
Files		Simple	x 7		
		Average	x 10		
		Complex	x 15		
File Total					
Enquiries		Simple	x 4		
		Average	x 5		
		Complex	x 6		
		Complex	x 7		
Enquiry Total					
Interface Files		Simple	x 5		
		Average	x 7		
		Complex	x 10		
Interface Total					
Total Unadjusted Function Point :					

Processing Complexity			
Factor	Value	Factor	Value
1. Data communications		8. On-line update	
2. Distributed function		9. Complex processing	
3. Performance		10. Code re-usability	
4. Heavily-used configuration		11. Conversion/installation ease	
5. Transaction rates		12. Operational ease	
6. On-line data entry		13. Mutliple site installation	
7. End user efficiency		14. Facilitate change	
Total Degree of Influence :			

Adjusted Function Point
Adjusted Function Point = Unadjusted Function Point x (0.65 + (0.01 x Total Degree of Influence))

Fig. A.4
Function Point Summary [Dreger, 1989]

Software Quality Agreement

Quality is Job 1
Ford commercial

As discussed in Chapters 5 and 9, the Software Quality Agreement should be negotiated, reviewed, and approved during the initial project planning session. It is essential that the quality requirements are defined for the project before any detailed estimation is undertaken. Any subsequent changes to the agreement including the ranking scores must be subject to the Change Control process described in Chapter 4.

Steps in Developing a Quality Agreement

- *Step 1—Evaluate and rank the stakeholders:* In conjunction with the team, the project manager should rank the project's stakeholders as essential land nonessential. Essential stakeholders are critical to the development or review of the project (usually Levels 4, 5, and 6); nonessential stakeholders must be consulted with but are not critical to the project's development (usually Levels 1, 2 and 3).
- *Step 2—Determine the project's quality requirements:* Using the model on page 172, determine in conjunction with the team which of the Quality Attributes are mandatory, nonmandatory, and not applicable. For each mandatory Quality Attribute, rank its related Quality Critera using the model contained in pages 173–175.
- *Step 3—Determine and review each stakeholder's ranking:* Preferably in a group session, interview each essential stakeholder, and review and determine his or her quality requirements using the same process as in Step 2.
- *Step 4—Derive final the ranking:* Evaluate all mandatory Quality Attributes looking for a majority agreement between the team and stakeholders (for example, if 80 percent of the stakeholders agree, then the attribute is mandatory for the project. Also evaluate the rankings for any potentially conflicting mandatory quality attributes. For example, Efficiency has a negative relationship with Maintainability, Flexibility, and Portability.
- *Step 5—Review the Quality Agreement with senior management:* The final rankings should be reviewed with senior management including the Steering Committee; any unresolved conflicts in the stakeholder's rankings should be raised for resolution by senior management.

While the rankings derived in this process are subjective, they provide a basis for discussing and defining what quality means for this product and for the product's stakeholders.

The use of the Quality Agreement as a basis for Quality Planning is discussed at the end of this appendix.

Quality Agreement and the Quality Plan

Most approaches to quality planning confuse the process quality and the product quality. For example, a typical quality planning technique is to use checklists to ensure that all relevant system development processes have been completed:

- Has the current system been analyzed?
- Has the data model been created?

And so on. While such an approach can ensure that the system development process has been followed, it does little to ensure that the development team clearly understands what specifically *is* the required product quality and that the quality requirements have not been compromised as the project moves through analysis, design, development, and implementation. Further, traditional quality planning does not ensure the quality of the quality assurance process; rather it ensures that the process has occurred.

The Quality Agreement provides a simple vehicle for specifying the quality requirements for a project. For example, assume that for System X, the quality attributes Conformity (data and function), Usability, and Job Impact are mandatory for the majority of key stakeholders. Based on this assumption, the project manager and team have a clear understanding of what aspects of the system must be reviewed during the development process.

The Quality Plan must include quality assurance processes that review each system deliverable from the perspective of the mandatory quality attributes first. The system analysis deliverables, such as data flow diagrams, data models, job specifications and human-computer interfaces, are assessed for Conformity, Usability, and Job Impact. As shown in Figure B.1, this assessment can be formalized into Baseline Reviews.

A Baseline Review is a Technical Review (see Freedman and Weinberg, op. cit.), in which a small group of expert peers reviews the system deliverables from the perspective of one of the mandatory quality attributes. In the example, there would be three Baseline reviews: a Conformity Review, a Usability Review, and a Job Impact Review. Typically, the review team changes for each Baseline Review. Work design experts would attend the Job Impact Review, while screen design, documentation, and usability experts would attend the Usability Review.

The Quality Agreement provides a framework for ensuring that the required quality is controlled through all the system development processes.

ATTRIBUTES	Criteria	Description
CONFORMITY	Completeness Correctness Traceability	Does the product have the desired data, function and procedures as required
USABILITY	Completeness Operability Support Training Correctness	Is the product easy to use, learn and understand from the user's perspective
EFFICIENCY	Processing Network Storage	Does the application use the hardware,system software and other resourcesefficiently
MAINTAINABILITY	Structure/Modularity Simplicity Commonality Documentation Self-descriptiveness	Is the system easy to maintain andcorrect
REUSABILITY	Self-descriptiveness Independance Structure/Modularity Commonality Application Independance Simplicity	Does the system use code and data that is capableof being used by other systems
FLEXIBILITY	Structure/Modularity Simplicity Documentation Self-descriptiveness	Is the system easy to enhance in order to add or modify function and data
RELIABILITY	Structure/Modularity Correctness Simplicity	Does the system operate without failure and with consistency
PORTABILITY	Independance Simplicity Structure/Modularity Self-descriptiveness	Is the system easy to migrate to another hardware, software environment
AUDITABILITY/SECURITY	Structure/Modularity Access Control Audit Control	Is the system secure from unauthorised access and is it auditable
JOB IMPACT	Correctness Work Dimensions Support Operability Training Network Documentation	Does the system provide acceptable working environment for direct users

adapted from McCall et al

Fig. B.1
Quality Agreement and Quality Plan

Software Quality Agreement
Project: Date: ...../...../.....

ATTRIBUTES	KEY STAKEHOLDER				
	Team				
CONFORMITY Does the product have the desired data, function and procedures arequired					
USABILITY Is the product easy to use, learn and understand from the user'sperspective					
EFFICIENCY Does the application use the hardware, system software and otherresources efficiently					
MAINTAINABILITY Is the system easy to maintain and correct					
REUSABILITY Does the system use code anddata that is capable of being used by other systems					
FLEXIBILITY Is the system easy to enhance in order to add or modify function anddata					
RELIABILITY Does the system operate without failure and with consistency					
PORTABILITY Is the system easy to migrate to another hardware, software environment					
AUDITABILITY/SECURITY Is the system secure from unauthorised access and is it auditable					
JOB IMPACT Does the system provide acceptable working environment for direct users					

M - MANDATORY N.A. - NOT APPLICABLE

Page ...... of ....

SYSTEM NAME : .... **DATE** ..../..../..../

MODULE NAME (Optional) : ....

Completeness

Low (-3)	-3 ... 0 ... +3	High (+3)
. minimum data requirements met as specified by clients . minimum functionality met	-3 0 +3	. total data requirements met as specified by clients and stakeholders . total functionality

Correctness

Low (-3)	-3 ... 0 ... +3	High (+3)
. data redundancy, incorrect data, corrupt data, poor file structure . high level of defects in code	-3 0 +3	. clean data, minimum data redundancy, structured files . zero code defects

Traceability

Low (-3)	-3 ... 0 ... +3	High (+3)
. no accurate analysis or design specs available - only module specs . no written requirements	-3 0 +3	. reviewed analysis, design and module specs available . all changes documented

Operability

Low (-3)	-3 ... 0 ... +3	High (+3)
. code and numeric interfaces . inadequate feedback on errors . cluttered screen design . inflexible screen navigation . inconsistent operation	-3 0 +3	. simple, human interface (eg icons, menus) . full error feedback and confirmation . well-designed screens, navigation . consistent operation

Support

Low (-3)	-3 ... 0 ... +3	High (+3)
. no user documentation, on-line help, etc . no help desk or expert contact person for queries	-3 0 +3	. full user documentation on-line help at all levels, well presented manuals . help-desk or support person full-time and expert

Training

Low (-3)	-3 ... 0 ... +3	High (+3)
no on-going education or computer-based education . no training manuals or case studies	-3 0 +3	. on-going and accessible education for system . training manuals and case studies available . on-line tutorials available and maintained

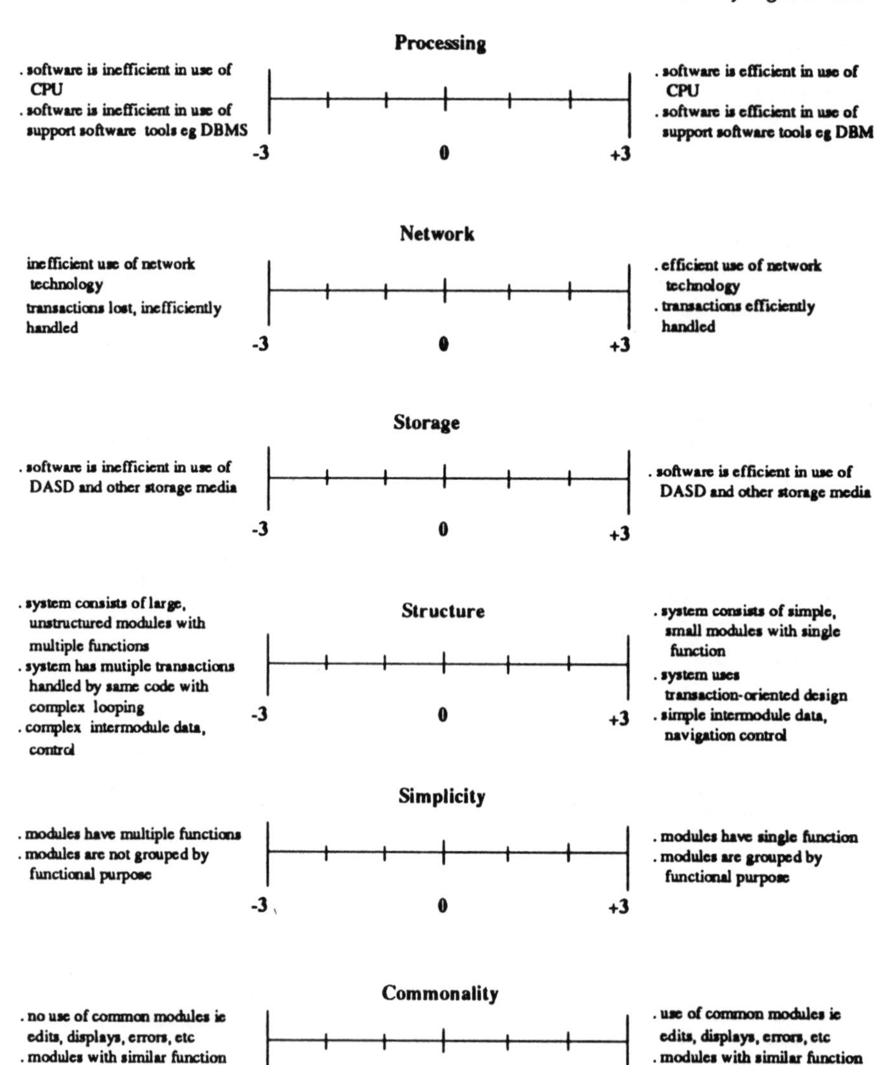
Processing
. software is inefficient in use of CPU
. software is inefficient in use of support software tools eg DBMS
. software is efficient in use of CPU
. software is efficient in use of support software tools eg DBM
-3
0
+3
Network
inefficient use of network technology
transactions lost, inefficiently handled
. efficient use of network technology
. transactions efficiently handled
-3
0
+3
Storage
. software is inefficient in use of DASD and other storage media
. software is efficient in use of DASD and other storage media
-3
0
+3
Structure
. system consists of large, unstructured modules with multiple functions
. system has mutiple transactions handled by same code with complex looping
. complex intermodule data, control
. system consists of simple, small modules with single function
. system uses transaction-oriented design
. simple intermodule data, navigation control
-3
0
+3
Simplicity
. modules have multiple functions
. modules are not grouped by functional purpose
. modules have single function
. modules are grouped by functional purpose
-3
0
+3
Commonality
. no use of common modules ie edits, displays, errors, etc
. modules with similar function are repeated
. use of common modules ie edits, displays, errors, etc
. modules with similar function are re-used
-3
0
+3

Documentation

. no system documentation
. no written specs
. no record of changes

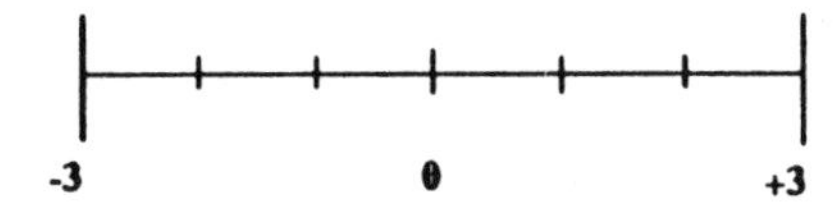

. full system documentation
. written specs
. changes recorded

Self-descriptiveness

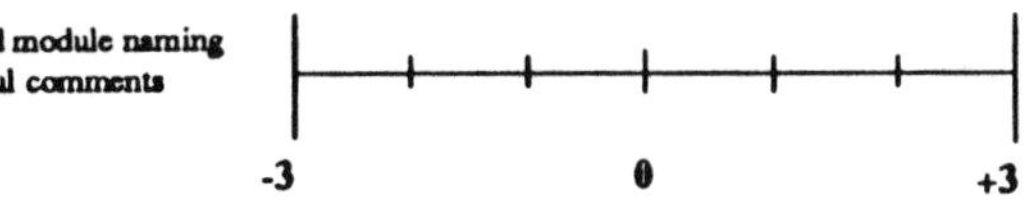

. accurate and meaningful data and module naming
. accurate comments

Independence

. software uses vendor-dependent languages or language extensions
. software is dependent on hardware and vendor-related support tools

-3 0 +3

. software does not use vendor-dependent languages or language extensions
. software is independent of hardware and vendor-related support tools

Application Independence

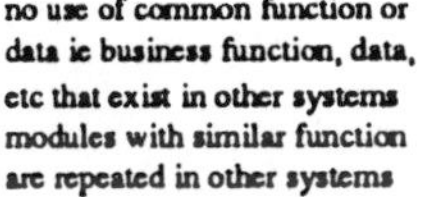

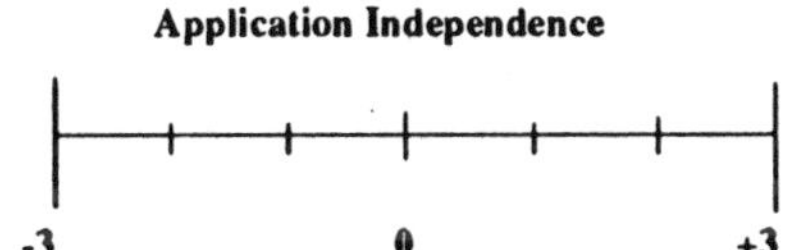

. use of common function, data ie business rules, data etc that exist in other systems
. modules with similar function are shared with other systems

Access Control

. no security protection or record of access
. system can be changed in production without review or authorisation

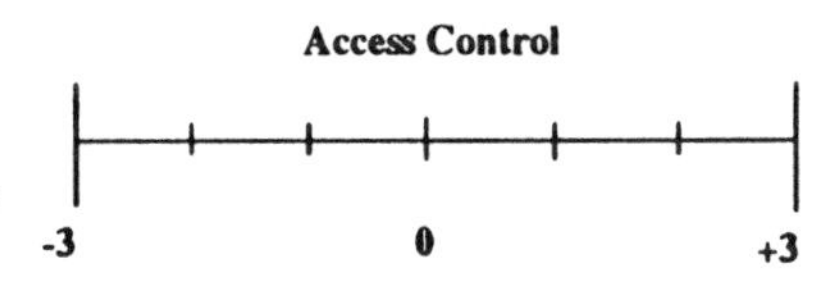

. full security protection or record of access
. system can't be changed in production without review or authorisation

Audit Control

. no use of audit trails, contol totals, system logs
. un-authorised changes can be made

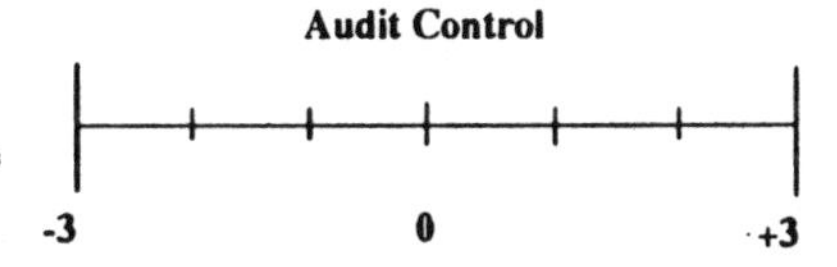

. full use of audit trails, control totals, system logs
. all changes are reviewed by Audit specialists

Work Dimensions

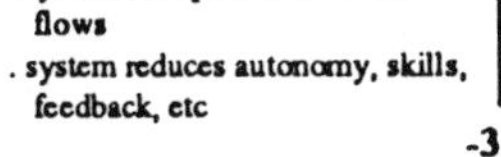

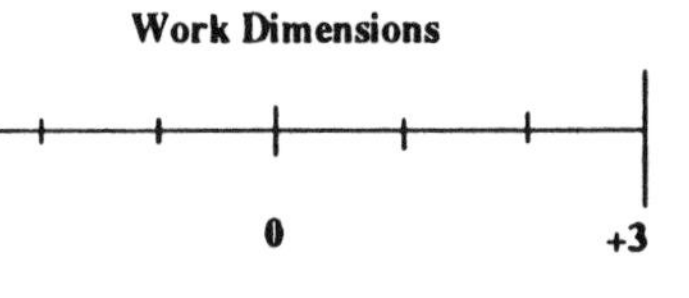

. system supports clerical work flows
. system enhances autonomy, skills, feedback, etc

B.5

The Essential Project Manager's Bookshelf

A book a day keeps the disasters away.

I have selected these books and references from my own personal library of over 300 professional books and from feedback from thousands of workshop participants. I have selected them based on four criteria:

1. They contain major or innovative statements on the project management process.
2. They are practical, not theoretical.
3. They provide detailed "hands-on" models, which can be used immediately.
4. They are interesting to read.

To keep the list manageable, I have excluded many essential general management and business books by people such as Tom Peters, Peter Drucker, Charles Handy, and so on, focusing on software project management only.

However, to undertake project management without access to those texts is similar to boxing with one hand behind your back.

Mary M. Parker, Robert J. Benson & H. E. Trainor, *Information Economics*, Englewood Cliffs, N.J.: Prentice Hall, 1988.

This book presents a rigorous and practical approach to project return on investment analysis that goes beyond the simple financial approaches available in most project management books. Using the popular concept of "added value," Parker, Benson and Trainor have a lot of innovative yet sensible ides on project justification and benefits modeling.

Tom Gilb, *Principles of Software Engineering Management. Reading, Mass.: Addison-Wesley, 1988.*

Tom is one of the great innovative thinkers of the industry. He presents a powerful approach to metrics—not just software—and this is interpersed with Tom's delightful lessons and laws of project management. A joy and challenge to read.

Tarke Abdel-Hamid & Stuart E. Madnick, *Software Project Dynamics*, Englewood Cliffs, N.J.: Prentice Hall, 1991.

This book is one of the major breakthroughs in third wave project management. Using theoretical yet practical models that they have been developing over the past 10 years (at least), Abdel-Hamid and Madnick provide a series of dynamic models showing the complex interactions of the major variables in project development. Although it is heavy going, it is a clear indication of the direction of project management practice and software.

Richard H. Thayer, ed., *Software Engineering Project Management.* Los Alamitos, Ca.; IEEE Computer Society Press, 1990.

This collection of papers represents the best of second wave project management concepts. It includes papers by Barry Boehm and other leading experts, and provides a good summary of the traditional project management process. Although the context of these papers is traditional, mechanistic processes, the tutorial provides a valuable background for understanding the more contemporary models in this handbook.

Donald C. Gause & Gerald M. Weinberg, *Exploring Requirements: Quality Before Design.* New York: Dorest House, 1989.

No essential collection is complete without at least one book by Gerry Weinberg. Along with Larry Constantine, Gerry is one of the few holistic observers of the computing industry. This book presents in Weinberg and Gause's witty style a challenging series of concepts and ideas on modeling system requirements, which is a vital supporting process for defining project scope and objectives.

G. Gordon Schulmeyer & James I. McManus, ed., *Handbook of Software Quality Assurance.* New York: Van Nostrand Reinhold, 1987.

There are so many books on software quality and quality in general that it is difficult to settle on one. The Schulmeyer and McManus book contains a number of excellent articles and papers and, given that we are all too busy to read, browsing through this book in a train, bus, or plane will introduce you to some of the more significant issues of software quality.

Bernard Londeix, *Cost Estimation for Software Development.* Reading, Mass.: Addison-Wesley, 1987.

This is not an easy book to get into, but it is worth the effort. Londeix provides a detailed and wide-ranging hands-on guide to the more esoteric estimation models such as Boehm's COCOMO, the Putnam-Norton model, and many others. A must for a serious understanding of alternative estimation processes.

Frederick P. Brooks, *The Mythical Man-Month.* Reading, Mass.: Addison-Wesley, 1975.

Written by Fred Brooks after his major learning experience on the OS/360 project, this short collection of essays is a true classic. Although the technically oriented essays are more of historical interest only, his management and project learnings are probably more relevant today than ever before. One read of this book could save your organization thousands of dollars!

T. Capers Jones, *Programming Productivity*.
New York: McGraw Hill, 1986.

When it comes to software metrics, Capers is simply the best. This book contains a vital set of project, software, and productivity metrics, which provide a virtual data base for deriving and supporting your estimates. In addition, the book contains pithy insights into the software industry. Also of interest is Capers' new book, *Applied Software Measurement* (New York: McGraw-Hill, 1991), which continues to refine some key metrics and provides a wide-ranging discussion on the state of the art of metrics.

Ed Yourdon & Toni Nash, eds., *American Programmer*.
New York: American Programmer Inc.

After many years as the guiding forces of the Structured Revolution, Ed and Toni have drawn on their extensive network of leading experts to create the most innovative monthly magazine for computing professionals. Unlike most professional magazines, which are filled with "press release quality" articles, each issue of *American Programmer* focuses on a specific topic such as project management, object orientation, and down-sizing using articles written by Larry Constantine, Meilir Page-Jones, Wayne Stevens, Chris Gane, Paul Ward, Rebecca Wirfs-Brock, and many other major computing people. It is a must for any computer professional interested in a nonvendor and independent view of the state-of-the-art.

Tom DeMarco and Tim Lister, eds., *Software State-of-the-art*.
New York: Dorest House Publishing, 1990.

As leading lights of Yourdon, Inc., Tom and Tim have used their eclectic views to select some fascinating papers. While the project management selection (with the exception of Tarek Abdel-Hamid's paper) is second wave, there are significant papers by Bill Curtis, Fred Brooks, Barry Boehm, Gerry Weinberg, and Warren McFarlan. Indeed a stellar cast and an interesting read.

Index

P

Q

R